AN ATLAS OF
MAMMALIAN
CHROMOSOMES

VOLUME 4

AN ATLAS OF
MAMMALIAN
CHROMOSOMES

VOLUME 4

T. C. HSU

Section of Cytology, Department
of Biology, The University of
Texas M. D. Anderson Hospital and
Tumor Institute, Houston, Texas

KURT BENIRSCHKE

Department of Pathology,
Dartmouth Medical School,
Hanover, New Hampshire

SPRINGER SCIENCE+BUSINESS MEDIA, LLC 1970

ISBN 978-1-4684-7386-5   ISBN 978-1-4615-6428-7 (eBook)
DOI 10.1007/978-1-4615-6428-7

© 1970 by Springer Science+Business Media New York
Originally published by Springer-Verlag New York Inc. in 1970
Softcover reprint of the hardcover 1st edition 1970

Title No. 3964

Introduction

It is our pleasure to see the Atlas of Mammalian Chromosomes now in its fourth volume. During the past few years, many reviews of the Atlas have appeared in many journals and in many languages. We appreciate the encouragements and compliments, and we shall try our best to continue this series. Volume five is almost ready.

Some criticisms and suggestions appeared in some reviews as well as in personal correspondence, and we would like to summarize and give answers or comments to them. We do not consider this to be an apologia, however.

1. *The quality of the photographs.*

    A. **The reproduction.** At the very beginning we decided that this Atlas series should be inexpensive enough so that students could afford it. Top quality would be most desirable, but the cost would be prohibitive. The current method used by Springer-Verlag, the dropout offset, definitely does not give optimal photographic reproduction. Naturally, if a book is as beautiful as it is useful, this is ideal; but when one must sacrifice, we prefer to sacrifice beauty for the sake of a reasonable price. We hope that the readers will find the reproduction of the karyotypes in the Atlas good enough for the intended purposes.

    B. **The picture itself.** Some of the photographs are not the best possible, however, we have tried to obtain the best available. We are particularly limited by the materials given to us by friends and, in this respect, we would like to enter a plea. Please supply us the best of your photographs with the chromosomes not overly condensed and with the least number of overlaps. Since we must fit all plates to the standard format, we request prints of the original metaphases or better still, the negatives.

    2. *Not enough representatives in each taxon for cytotaxonomy studies.* It must be remembered that each volume contains only 50 species. If we make an attempt to present one taxon of mammals exhaustively to make a meaningful phylogenetic comparison, other taxa will have to be sacrificed accordingly. By adding a few species of each taxon in every volume, the Atlas will probably make more persons happy, a practice that will continue.

    3. *Autoradiographs accompanying karyotypes undesirable.* We no longer include autoradiographs in the Atlas.

    4. *No pictures of the animals.* We cannot comply with some readers' wishes. We are exhausted just doing the chromosomes and the expense is prohibitive.

5. *Not enough representatives other than species from North America.* Both of us reside in the United States, and it is, thus, natural that we have easier access to U.S. mammals than those from other continents. We have tried to include, whenever possible, species from various lands and welcome contributions by cytologists of all countries to make this series a real service.

6. *No locality data.* We include locality data whenever they are available. It is often impossible, however, to obtain accurate locality information from zoo animals or animals purchased from dealers. We believe that it is worse to give false locality data than none at all. We urge that friends who contribute karyotypes furnish us with reliable records.

7. *No definition for chromosome morphology.* The main purpose of the Atlas is to present pictorial descriptions of mammalian karyotypes rather than written descriptions that can never be adequate. We use such terms as metacentrics, submetacentrics, etc., for convenience. Indeed, there is no nomenclature system for chromosome morphology that is internationally accepted. Thus, a submetacentric chromosome of one cytologist may be called subtelocentric by another, and a subtelocentric of one may be called acrocentric by another. All cytologists may have experienced a change of mind about how to call a particular chromosome. This is especially true when chromosomes with different degrees of condensation are compared. A relatively short second arm observed at early metaphase may become a mere knob-like structure at full metaphase, particularly if the cell has been arrested by colchicine or vinblastine for sometime. Thus, the same chromosome may be called subtelocentric or acrocentric in different cells.

Probably the most reasonable nomenclature proposal is that of Levan, Fredga and Sandberg (HEREDITAS, 1964); but in order to describe the chromosomes accurately according to Levan *et al.*, one needs to construct an idiogram for each species, which means measurements of all chromosomes from a number of good metaphase plates. This is highly impractical even though it would be a desirable feature.

The readers may have noticed that we have made some revision in our nomenclature system. Briefly, they can be described as follows:

Metacentric chromosome—two arms more or less equal.

Submetacentric chromosome—two arms unequal in length, but not so unequal that the long arm is three or more times as long as the short. This type of definition is not always satisfactory, because the absolute length of a particular chromosome may give an illusive impression when the chromosomes are visually inspected instead of measured.

Subtelocentric chromosome—the long arm of a chromosome is at least three or four times as long as the short. Again this may be illusive. For example, the same proportion, say 1:4, may give a subtelocentric appearance in a relatively short chromosome, but a submetacentric appearance in a long chromosome.

Acrocentric chromosome—the chromosome either appears to be monoarmed or possesses only a knob-like short arm.

Admittedly, the foregoing system is crude. Without detailed analysis on many karyotypes, it is dangerous to define each chromosome without reservation. Thus, our descriptions are only approximate. When one works with a particular species or a group of species, he must construct idiogram(s) for his critical research work.

8. *No zoologist's name under the species name.* We discussed this point in some detail at the beginning and decided not to follow it. Our main point is that one can always copy from one book or another the zoologist's name and, thereby, give the identification some false authenticity. In an occasional case species identification may eventually turn out to be wrong, and the attachment of a zoologist's name to the species name, therefore, does not provide any more security. For North American species, one can always find the complete information in Hall and Kelson.

Naturally, we hope to continue to improve the quality and the format of this Atlas and welcome constructive criticisms. We would like to learn from our readers whether they desire to have representations of variability of a particular species. For example, there are many different karyotypes in practically every species of pocket gophers, genus *Thomomys*, and it is not reasonable to select any one as the typical karyotype. However, in order to present variability, a number of plates may be required for a single species.

January, 1970   T. C. Hsu, K. Benirschke

Contents, Volume 4

CALLITHRICIDAE

COLOBIDAE

PONGIDAE

Cumulative Contents
(Volumes 1 to 4)

New References for Previous Volumes

In Volume 4 there are 50 new species of various Orders. These should be integrated with the Folios of Volumes 1, 2 and 3. A new cumulative index is supplied, hence the old index should be discarded.

In addition to the new Folios, 10 pages of references for the Folios of Volumes 1, 2 and 3 are prepared to keep the Atlas up-to-date and useful. These references are arranged in such a fashion that they can be cut out and pasted into the previous Folios in sequence.

MARSUPIALIA

Vol. 3, Folio 101

4) Hayman, D. L. and Martin, P. G.: Cytogenetics of marsupials. In
Comparative Mammalian Cytogenetics (Benirschke, K., ed.), Springer-Verlag,
New York, 1969.

Vol. 3, Folio 102

2) Hayman, D. L. and Martin, P. G.: Cytogenetics of marsupials. In
Comparative Mammalian Cytogenetics (Benirschke, K., ed.), Springer-Verlag,
New York, 1969.

Vol. 1, Folio 1

9) Goh, K. O.: Inner chromosomal structures in the somatic metaphase of
rat kangaroo (Potorous tridactylis). A light microscopic observation.
Cytologia 32:416, 1967.

10) Sinha, A. K.: Somatic association of homologous chromosomes. Proc.
XII Intern. Congress Genet. Vol. I, 163, 1968.

11) Hayman, D. L. and Martin, P. G.: Cytogenetics of marsupials. In
Comparative Mammalian Cytogenetics (Benirschke, K., ed.), Springer-Verlag,
New York, 1969.

Vol. 2, Folio 51

9) Steyer, L. C. and Paulete-Vanrell, J.: Análise da cromatina sexual em
Didelphis paraguayensis Oken, 1816 (Marsupialia). Pesquisas No. 22
(Instituto Anchietano de Pesquisas, Brasil) 1969.

INSECTIVORA

Vol. 2, Folio 52

7) Citoler, P. and Gropp, A.: DNS-Replikation von autosomalem
Heterochromatin. Exp. Cell Res. 54:337, 1969.

8) Gropp, A.: Cytologic mechanisms of karyotype evolution in insectivores.
In Comparative Mammalian Cytogenetics (Benirschke, K., ed.), Springer-Verlag,
New York, 1969.

9) Gropp, A. and Citoler, P.: Patterns of autosomal heterochromatin. In
Comparative Mammalian Cytogenetics (Benirschke, K., ed.), Springer-Verlag,
New York, 1969.

INSECTIVORA

Vol. 2, Folio 53

7) Gropp, A.: Cytologic mechanisms of karyotype evolution in insectivores. In Comparative Mammalian Cytogenetics (Benirschke, K., ed.), Springer-Verlag, New York, 1969.

8) Gropp, A. and Citoler, P.: Patterns of autosomal heterochromatin. In Comparative Mammalian Cytogenetics (Benirschke, K., ed.), Springer-Verlag, New York, 1969.

Vol. 2, Folio 54

3) Gropp, A.: Cytologic mechanisms of karyotype evolution in insectivores. In Comparative Mammalian Cytogenetics (Benirschke, K., ed.), Springer-Verlag, New York, 1969.

CHIROPTERA

Vol. 3, Folio 104

3) Beçak, M. L., Batistic, R. F., Vizotto, L. D. and Beçak, W.: Sex determining mechanism XY_1Y_2 in Artibeus lituratus lituratus (Chiroptera-Phyllostomidae). Experientia 25:81, 1969.

LAGOMORPHA

Vol. 1, Folio 8

15) Valenti, C. and Friedman, E. A.: Long-term cultivation of diploid rabbit skin cells. Texas Rep. Biol. Med. 26:336, 1968.

16) Pauly, J. L., Caron, G. A. and Suskind, R. R.: Blast transformation of lymphocytes from guinea pigs, rats, and rabbits induced by mercuric chloride in vitro. J. Cell Biol. 40:847, 1969.

17) Chang, M. D., Pickworth, S. and McGaughey, R. W.: Experimental hybridization and chromosomes of hybrids. In Comparative Mammalian Cytogenetics (Benirschke, K., ed.), Springer-Verlag, New York, 1969.

RODENTIA

Vol. 2, Folio 73

7) Pauly, J. L., Caron, G. A. and Suskind, R. R.: Blast transformation of lymphocytes from guinea pigs, rats, and rabbits induced by mercuric chloride in vitro. J. Cell Biol. 40:847, 1969.

Vol. 1, Folio 13

11) Zakharov, A. F. and Egolina, N. A.: Asynchrony of DNA replication and mitotic spiralization along heterochromatic portions of Chinese hamster chromosomes. Chromosoma 23:365, 1968.

12) Schmid, W. and Leppert, M. F.: Rates of DNA synthesis in heterochromatic and euchromatic segments of the chromosome complements of two rodents. Cytogenetics 8:125, 1969.

13) Ford, C. E.: Meiosis in mammals. In Comparative Mammalian Cytogenetics (Benirschke, K., ed.), Springer-Verlag, New York, 1969.

14) Fraccaro, M., Gustavsson, I., Hultén, M., Lindsten, J. and Tiepolo, L.: Late-replicating Y chromosome in spermatogonia of the Chinese hamster (Cricetulus griseus). Cytogenetics 8:263, 1969.

Vol. 1, Folio 14

7) Hill, R. N. and Yunis, J. J.: Mammalian X-chromosomes: Change in patterns of DNA replication during embryogenesis. Science 155:1120, 1967.

8) Clendenin, T. M.: Intraperitoneal colchicine and hypotonic KCl for enhancement of abundance and quality of meiotic chromosome spreads from hamster testis. Stain Technol. 44:63, 1969.

9) Lemon, J. G. and Morton, W. R. M.: Oogenesis in the golden hamster (Mesocricetus auratus). A study of the first meiotic prophase. Cytogenetics 7:376, 1968.

10) Mukherjee, B. B. and Ghosal, S. G.: Replicative differentiation of mammalian sex-chromosomes during spermatogenesis. Exp. Cell Res. 54:101, 1969.

11) Emmons, L. R. and Husted, L.: The sex bivalent of the golden hamster. J. Hered. 53:227, 1962.

12) Galton, M. and Holt, S.: Culture of peripheral blood leucocytes of the golden hamster. Proc. Soc. Exp. Biol. Med. 114:218, 1963.

13) Matthey, R.: Les chromosomes de Mesocricetus auratus Waterh. Rev. Suisse Zool. 60:466, 1953.

14) Walknowska, J.: The chromosomes in ontogenesis of golden hamster (Mesocricetus auratus). Folia Biol. (Krakow) 12:321, 1964.

Vol. 1, Folio 15

1) Schmid, W. and Leppert, M. F.: Karyotyp, Heterochromatin und DNS-Werte bei 13 Arten von Wühlmäusen (<u>Microtinae</u>, <u>Mammalia</u>-<u>Rodentia</u>). Arch. Julius Klaus-Stiftung <u>43</u>:88, 1968.

Vol. 1, Folio 17

16) Cattanach, B. M. and Pollard, C. E.: An XYY sex-chromosome constitution in the mouse. Cytogenetics <u>8</u>:80, 1969.

17) Mukherjee, B. B. and Ghosal, G.: Replicative differentiation of mammalian sex-chromosomes during spermatogenesis. Exp. Cell Res. <u>54</u>:101, 1969.

18) Ford, C. E.: Meiosis in mammals. In <u>Comparative Mammalian Cytogenetics</u> (Benirschke, K., ed.), Springer-Verlag, New York, 1969.

19) Gropp, A.: Cytologic mechanisms of karyotype evolution in insectivores. In <u>Comparative Mammalian Cytogenetics</u> (Benirschke, K., ed.), Springer-Verlag, New York, 1969.

20) Nadler, C. F.: Chromosomal evolution in rodents. In <u>Comparative Mammalian Cytogenetics</u> (Benirschke, K., ed.), Springer-Verlag, New York, 1969.

Vol. 1, Folio 18

18) Pauly, J. L., Caron, G. A. and Suskind, R. R.: Blast transformation of lymphocytes from guinea pigs, rats, and rabbits induced by mercuric chloride <u>in vitro</u>. J. Cell Biol. <u>40</u>:847, 1969.

Vol. 3, Folio 112

8) Schmid, W. and Leppert, M. F.: Karyotyp, Heterochromatin und DNS-Werte bei 13 Arten von Wühlmäusen (<u>Microtinae</u>, <u>Mammalia</u>-<u>Rodentia</u>). Arch. Julius Klaus-Stiftung <u>43</u>:88, 1968.

Vol. 3, Folio 118

1) Hsu, T. C.: Robertsonian fusion between homologous chromosomes in a natural population of the least cotton rat, <u>Sigmodon minimus</u> (Rodentia, Cricetidae). Experientia <u>25</u>:205, 1969.

Vol. 2, Folio 70

4) Schmid, W. and Leppert, M. F.: Karyotyp, Heterochromatin und DNS-Werte bei 13 Arten von Wühlmäusen (Microtinae, Mammalia-Rodentia). Arch. Julius Klaus-Stiftung 43:88, 1968.

Vol. 3, Folio 120

3) Schmid, W. and Leppert, M. F.: Karyotyp, Heterochromatin und DNS-Werte bei 13 Arten von Wühlmäusen (Microtinae, Mammalia-Rodentia). Arch. Julius Klaus-Stiftung 43:88, 1968.

Vol. 2, Folio 69

8) Schmid, W. and Leppert, M. F.: Rates of DNA synthesis in heterochromatic and euchromatic segments of the chromosome complements of two rodents. Cytogenetics 8:125, 1969.

9) Schwarzacher, H. G. and Pera, F.: Multipolar mitosis and somatic segregation in cell cultures of Microtus agrestis. In Comparative Mammalian Cytogenetics (Benirschke, K., ed.), Springer-Verlag, New York, 1969.

10) Schmid, W. and Leppert, M. F.: Karyotyp, Heterochromatin und DNS-Werte bei 13 Arten von Wühlmäusen (Microtinae, Mammalia-Rodentia). Arch. Julius Klaus-Stiftung 43:88, 1968.

11) Pera, F. and Schwarzacher, H. G.: Die Verteilung der Chromosomen auf die Tochterzellkerne multipolarer Mitosen in euploiden Gewebekulturen von Microtus agrestis. Chromosoma 26:337, 1969.

Vol. 3, Folio 113

1) Baker, R. J. and Mascarello, J. T.: Karyotypic analyses of the genus Neotoma (Cricetidae, Rodentia). Cytogenetics 8:187, 1969.

Vol. 2, Folio 62

1) Baker, R. J. and Mascarello, J. T.: Karyotypic analyses of the genus Neotoma (Cricetidae, Rodentia). Cytogenetics 8:187, 1969.

Vol. 2, Folio 61

1) Baker, R. J. and Mascarello, J. T.: Karyotypic analyses of the genus Neotoma (Cricetidae, Rodentia). Cytogenetics 8:187, 1969.

Vol. 3, Folio 130

2) Wurster, D. H.: Cytogenetic and phylogenetic studies in carnivora.
In Comparative Mammalian Cytogenetics (Benirschke, K., ed.), Springer-Verlag,
New York, 1969.

Vol. 1, Folio 32

2) Wurster, D. H.: Cytogenetic and phylogenetic studies in carnivora.
In Comparative Mammalian Cytogenetics (Benirschke, K., ed.), Springer-Verlag,
New York, 1969.

Vol. 1, Folio 27

8) Chang, M. D., Pickworth, S. and McGaughey, R. W.: Experimental
hybridization and chromosomes of hybrids. In Comparative Mammalian
Cytogenetics (Benirschke, K., ed.), Springer-Verlag, New York, 1969.

Vol. 1, Folio 20

15) Tzessarskaya, T. P.: Somatic chromosomes of the dog. Genetika 10:158,
1968.

Vol. 1, Folio 31

12) Jones, T. C.: Anomalies of sex chromosomes in tortoiseshell male cats.
In Comparative Mammalian Cytogenetics (Benirschke, K., ed.), Springer-Verlag,
New York, 1969.

Vol. 2, Folio 81

8) Evsikov, V. I. and Isakova, G. K.: Some results of karyological studies
in the minks (Lutreola vison Brisson) of various genotypes. Genetika 4:34,
1968.

9) Chang, M. D., Pickworth, S. and McGaughey, R. W.: Experimental
hybridization and chromosomes of hybrids. In Comparative Mammalian
Cytogenetics (Benirschke, K., ed.), Springer-Verlag, New York, 1969.

10) Itoh, M., Sasaki, M., Shinba, H. and Shiota, Y.: The chromosomes of
four mutant strains of the mink, Mustela vison (Carnivora, Mustelidae).
Zool. Magaz. (Dobutsugaku Zasshi) 77:374, 1968.

CARNIVORA

Vol. 3, Folio 123

3) Ulbrich, F. and Schmitt, J.: Die Chromosomen des Erdwolfs Proteles cristatus (Sparrmann, 1783). Z. Säugetierk. 34:61, 1969.

Vol. 1, Folio 28

2) Hsu, T. C. and Mead, R. A.: Mechanisms of chromosomal changes in mammalian speciation. In Comparative Mammalian Cytogenetics. (Benirschke, K., ed.), Springer-Verlag, New York, 1969.

ARTIODACTYLA

Vol. 1, Folio 44

19) Mukherjee, B. B., Wright, W. C., Ghosal, S. K., Burkholder, G. D. and Mann, K. E.: Further evidence for the simultaneous initiation of DNA replication in both X chromosomes of bovine females. Nature 220:714, 1968.

20) Bambhani, R. and Kuspira, J.: Chromosome preparations of bovine leucocytes. Experientia 25:83, 1969.

21) Gustavsson, I., Fraccaro, M., Tiepolo, L. and Lindsten, J.: Presumptive X-autosome translocation in a cow: preferential inactivation of the normal X chromosome. Nature 218:183, 1968.

22) Popesco, P. C.: Observations cytogénétiques chez les bovins charolais normaux et culards. Ann. Génét. 11:262, 1968.

23) Popesco, P. C.: Tehnica de microcultura din singe periferic pentru studiul cromozomilor la animale domestice. St. si Cerc. Biol. Ser. Zool. 20:421, 1968.

24) Herzog, R. and Steffensen, D. M.: The pattern of protein synthesis in late S and G_2 of bovine sex chromosomes. Cytogenetics 7:471, 1968.

Vol. 2, Folio 92

15) DeGrouchy, J., Lauvergne, J. J. and Ricordeau, G.: Etudes cytogénétiques chez 16 chevres intersexuées. C. R. Acad. Sc. (Paris) 260:2932, 1965.

16) Schmitt, J. and Ulbrich, F.: Die Chromosomen verschiedener Caprini. Simpson, 1945. Z. Säugetierk. 33:180, 1968.

Vol. 1, Folio 45

16) Ilberry, P. L. T., Alexander, G. and Williams, D.: The chromosomes of sheep x goat hybrids. Austr. J. Biol. Sc. 20:1245, 1967.

17) Schmitt, J. and Ulbrich, F.: Die Chromosomen verschiedener Caprini. Simpson, 1945. Z. Säugetierk. 33:180, 1968.

18) Popesco, P. C.: Tehnica de microcultura din singe periferic pentru studiul cromozomilor la animale domestice. St. si Cerc. Biol. Ser. Zool. 20: 421, 1968.

Vol. 1, Folio 38

19) Cornefert-Jensen Fr., Hare, W. C. D. and Abt, D. A.: Identification of the sex chromosomes of the domestic pig. J. Hered. 59:251, 1968.

20) Breeuwsma, A. J.: A case of XXY sex chromosome constitution of an intersex pig. J. Reprod. Fert. 16:119, 1968.

21) Harvey, M. J. A.: A male pig with an XXY/XXXY sex chromosome complement. J. Reprod. Fert. 17:319, 1968.

22) Ruddle, F. H.: Quantitation and automation of chromosomal data with special reference to the chromosomes of the Hampshire pig (Sus scrofa). In Cytogenetics of Cells in Culture (Harris, R. J. C., ed.), Academic Press, New York, 1964.

23) McFee, A. F. and Banner, M. W.: Inheritance of chromosome number in pigs. J. Reprod. Fert. 18:9, 1969.

24) Antonio, E. di: Il cariogramma del suino. Vet. ital. 15:925, 1964.

Vol. 3, Folio 134

4) Herzog, A. and Höhn, H.: Darstellung der Chromosomen aus Knochenmark-szellen beim Reh (Capreolus capreolus) und Rottier (Cervus elaphus). Z. Jagdwiss. 13:118, 1967.

Vol. 1, Folio 39

3) Gropp, A., Giers, D. and Tettenborn, U.: Das Chromosomenkomplement des Wildschweins (Sus scrofa). Experientia 25:778, 1969.

ARTIODACTYLA

Vol. 3, Folio 135

3) Gustavsson, I. and Sundt, C. O.: A note on the somatic Y chromosome of reindeer (Rangifer tarandus, L.). Acta vet. scand. 10:44, 1969.

Vol. 2, Folio 86

5) Herzog, A. and Höhn, H.: Darstellung der Chromosomen aus Knochenmark- szellen beim Reh (Capreolus capreolus) und Rottier (Cervus elaphus). Z. Jagdwiss. 13:118, 1967.

PRIMATES

Vol. 1, Folio 46

7) Egozcue, J., Perkins, E. M. and Hagemenas, F.: Chromosomal evolution in marmosets, tamarins, and pinchés. Folia primat. 9:81, 1968.

Vol. 1, Folio 47

5) Egozcue, J., Perkins, E. G. and Hagemenas, F.: Chromosomal evolution in marmosets, tamarins, and pinchés. Folia primat. 9:81, 1968.

Vol. 1, Folio 48

9) Chiarelli, B.: Chromosome polymorphism in the species of the genus Cercopithecus. Cytologia 33:1, 1968.

Vol. 1, Folio 50

10) Carr, D. H.: Chromosomal abnormalities in clinical medicine. In Progress in Medical Genetics, Vol. VI (Steinberg, A. G. and Bearn, A. G. eds.), Grune and Stratton, New York, 1969.

Vol. 3, Folio 147

9) Egozcue, J.: Cytological evidence suggestive of crossing over within the mammalian X chromosome. Experientia 24:1275, 1968.

10) Egozcue, J.: Primates. In Comparative Mammalian Cytogenetics (Benirschke, K., ed.), Springer-Verlag, New York, 1969.

PRIMATES

Vol. 2, Folio 99

3) Egozcue, J., Perkins, E. M. and Hagemenas, F.: Chromosomal evolution
in marmosets, tamarins, and pinchés. Folia primat. $\underline{9}$:81, 1968.

Vol. 2, Folio 100

6) Egozcue, J., Perkins, E. M. and Hagemenas, F.: Chromosomal evolution
in marmosets, tamarins, and pinches. Folia primat. $\underline{9}$:81, 1968.

Vol. 3, Folio 146

3) Arrighi, F. E., Sorenson, M. W. and Shirley, L. S.: Chromosomes of the
tree shrews (Tupaiidae). Cytogenetics $\underline{8}$:199, 1969.

Vol. 2, Folio 97

6) Arrighi, F. E., Sorenson, J. W. and Shirley, L. S.: Chromosomes of the
tree shrews (Tupaiidae). Cytogenetics $\underline{8}$:199, 1969.

Vol. 3, Folio 149

1) Benirschke, K.: Cytogenetic contributions to primatology. In Experimental
Medicine and Surgery in Primates. Goldsmith, Edward I. and Moor-Jankowski, J.,
eds. Ann. N. Y. Acad. Sci. $\underline{162}$:217, 1969.

Order: MARSUPIALIA

Family: PERAMELIDAE

Perameles nasuta (Long-nosed bandicoot)
$2n = 14$

<u>Perameles nasuta</u> (Long-nosed bandicoot)

2n=14

AUTOSOMES : 12 Metacentrics and submetacentrics

SEX CHROMOSOMES: X Metacentric
                   Y Acrocentric

All chromosomes are morphologically distinguishable. The X chromosome has a secondary constriction on the long arm. The sixth autosomal pair is satellited.

The karyotypes presented here are gifts of Dr. Laird G. Jackson, Jefferson Medical School, Philadelphia, Pennsylvania, USA. The animals were trapped in the vicinity of Sydney, N.S.W., Australia. Skin tissue culture cells were used for cytological preparations from pouch young.
Hayman and Martin describe the sex chromosome mosaicism that seems to occur in this species. Testes have 14 XY, ovaries 14 XX. Bone marrow, lymphocytes and spleen have 13 XO in both sexes. Corneal epithelium and intestine have 14 XY or 14 XX, and Jackson and Ellem find 2n=14 in the skin cultures of two males and four females they studied. They also provide arm ratio measurements.

REFERENCES:

1) Sharman,G.B.: The mitotic chromosomes of marsupials and their bearing on taxonomy and phylogeny. Austr. J. Zool. <u>9</u>:38, 1961.

2) Hayman,D.L. and Martin, P.G.: Sex chromosome mosaicism in the marsupial genera Isoodon and Perameles. Genetics <u>52</u>:1201, 1965.

3) Jackson, L.G. and Ellem, K.A.O.: The karyotype of the Australian long-nosed bandicoot (<u>Perameles</u> <u>nasuta</u>). Cytogenetics <u>7</u>:183, 1968.

4) Hayman, D.L. and Martin, P.G.: Cytogenetics of marsupials. In "<u>Comparative</u> <u>Mammalian</u> <u>Cytogenetics</u>"(Benirschke,K., ed.), Springer-Verlag, New York, 1969.

5) Walton, S.: Sex chromatin in an Australian marsupial <u>Perameles</u> <u>nasuta</u> Geoffroy, 1804. Experientia <u>25</u>:535, 1969.

Perameles nasuta (Long-nosed bandicoot)
2n=14

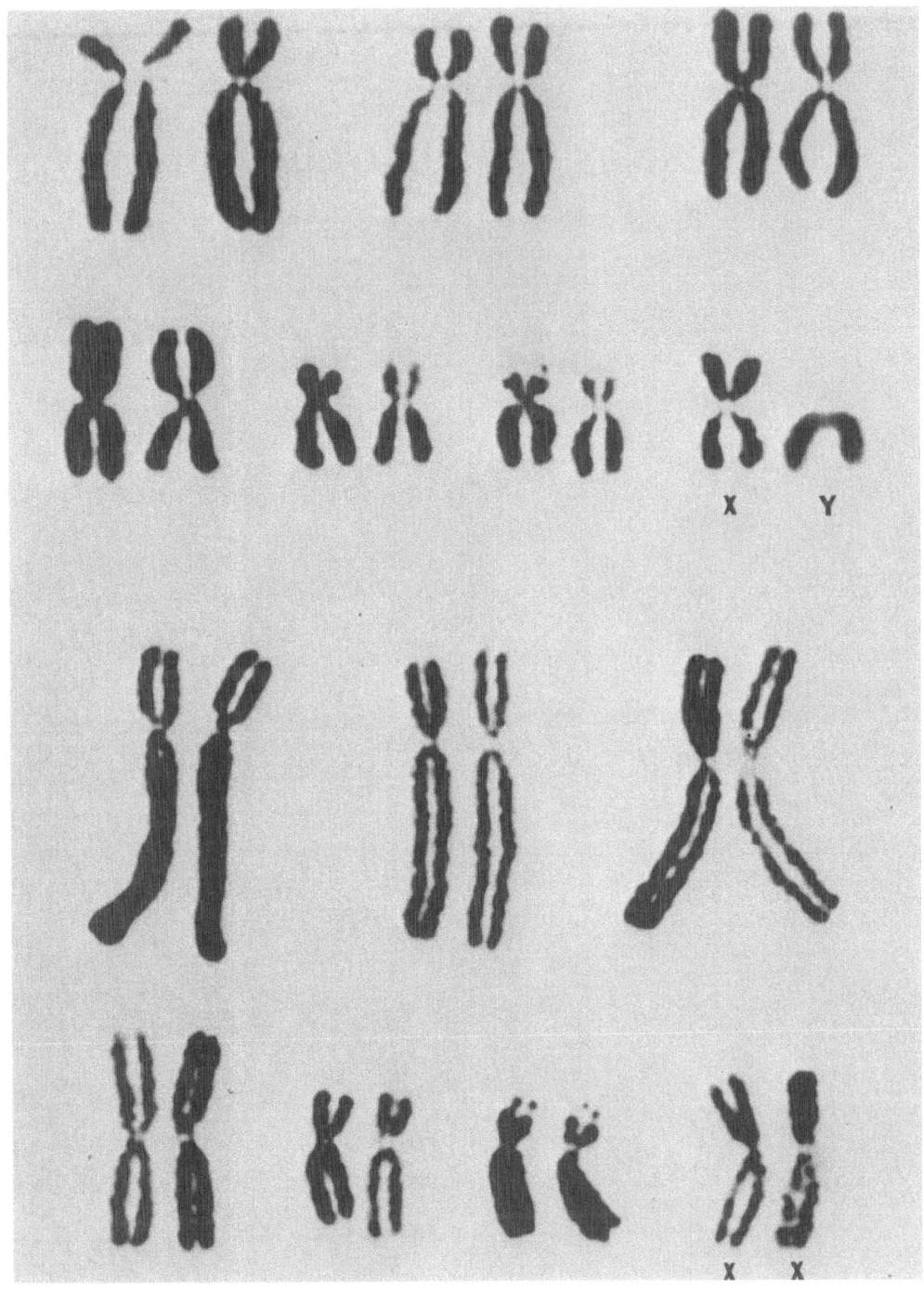

Order: MARSUPIALIA

Family: MACROPODIDAE

Wallabia (Protemnodon) bicolor (Black-tailed wallaby)

$2n = 11\,\sigma,\ 10\,♀$

Wallabia (Protemnodon) bicolor (Black-tailed wallaby)

$2n=11\sigma$, $10\female$

AUTOSOMES: 8 Metacentrics and submetacentrics
            2 Acrocentrics with translocated X

SEX CHROMOSOMES: X Acrocentric, translocated to large acrocentric autosome
                  Y Minute

All chromosome pairs are morphologically distinguishable. The X chromosome, approximately 5% of the haploid set, is a small acrocentric that is translocated onto the large acrocentric autosome. This interpretation differs from the opinions published of a Y_1Y_2 system but is in conformity with other mammals having such a sex chromosomal arrangement.

The karyotypes came from leukocyte cultures and were kindly donated by Dr. David L. Hayman, University of Adelaide, Australia.

REFERENCES:

1) Sharman, G.B.: The mitotic chromosomes of marsupials and their bearing on taxonomy and phylogeny. Austral. J. Zool. 9: 38, 1961.

2) Moore, R.C. and Gregory, G.: Biometrics of the karyotype of Protemnodon bicolor, with reference to the limitations in accuracy of identifying human chromosomes. Nature 200: 234, 1963.

3) Hayman, D.L. and Martin, P.G.: An autoradiographic study of DNA synthesis in the sex chromosomes of two marsupials with an XX/XY_1Y_2 sex chromosome mechanism. Cytogenetics 4: 209, 1965.

4) Hayman, D.L. and Martin, P.G.: Cytogenetics of marsupials. In Comparative Mammalian Cytogenetics (Benirschke, K., ed.), Springer-Verlag, N. Y., 1969.

Wallabia (Protemnodon) bicolor (Black-tailed wallaby)

2n=11♂, 10♀

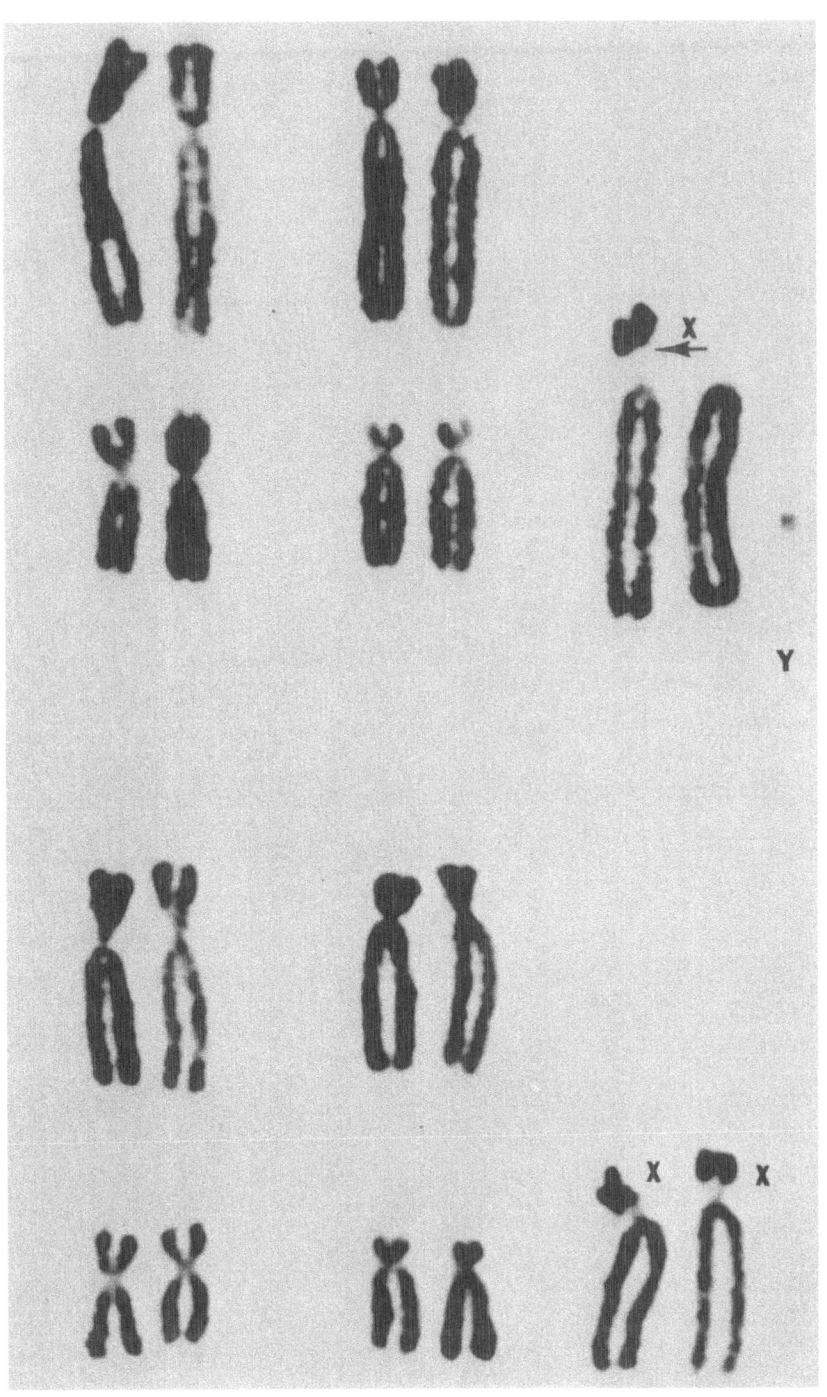

Order: INSECTIVORA

Family: TENRECIDAE

Centetes ecaudatus
$2n = 38$

Centetes ecaudatus

2n=38

AUTOSOMES: 36 Metacentrics, submetacentrics and subtelocentrics

SEX CHROMOSOMES: X Submetacentric
                  Y Minute

　　　Several pairs of large chromosomes have highly unequal arms.
Identification of the X chromosome is equivocal because several medium-sized
autosomes have similar morphology. Identification of the Y is unequivocal.

　　　The karyotypes are gifts of Dr. D. S. Borgoankar, Johns Hopkins Hospital,
Baltimore, Maryland, USA. The specimens were collected in Madagascar. Bone
marrows were used for cytological preparations.

REFERENCES:

1) Borgaonkar, D.S.: Insectivora Cytogenetics. In Comparative Mammalian
Cytogenetics (Benirschke, K., ed.), Springer-Verlag, New York, 1969.

Order: INSECTIVORA

Family: TENRECIDAE

Centetes ecaudatus

2n=38

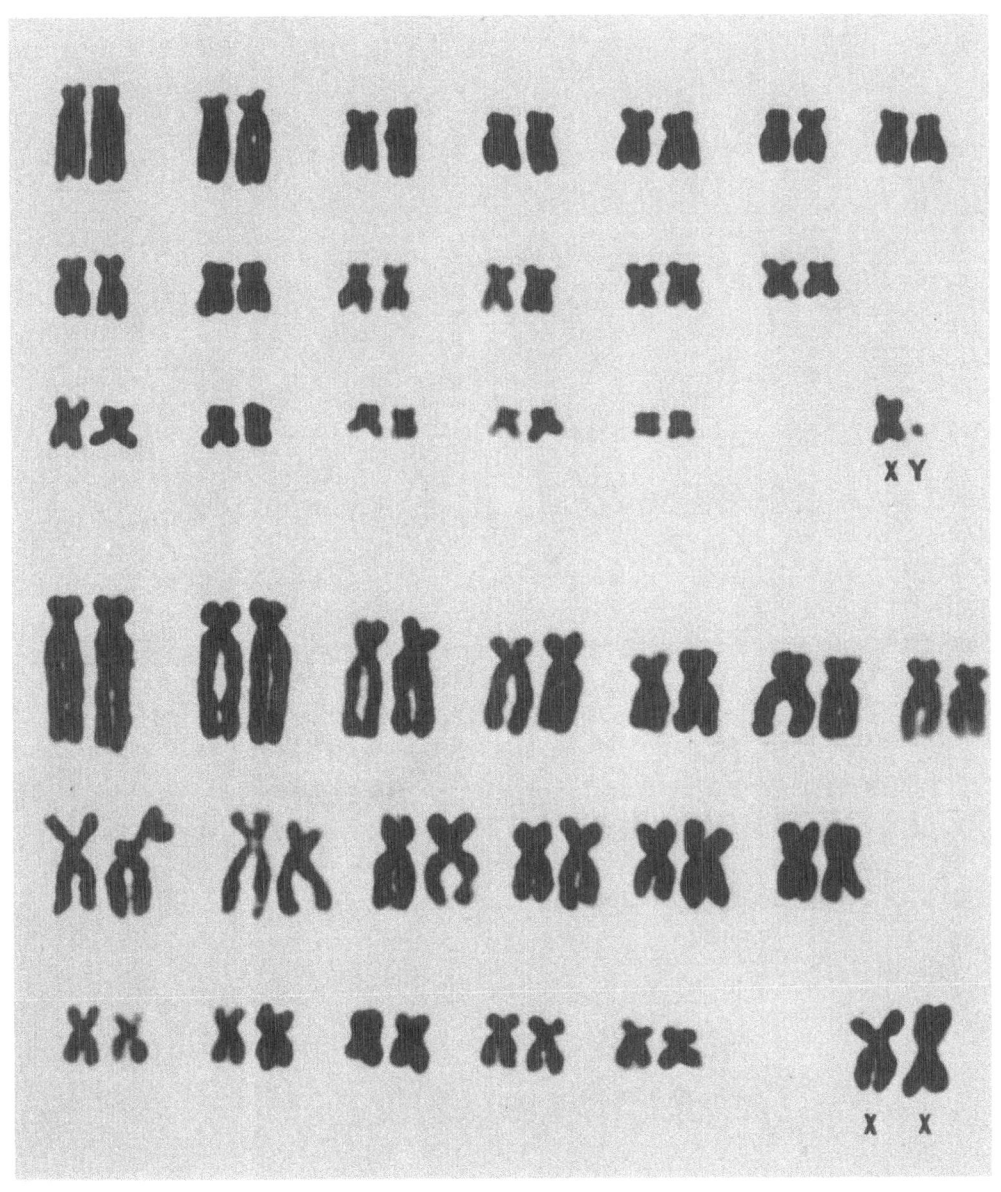

Order: INSECTIVORA

Family: TENRECIDAE

Microgale dobsoni
2n = 30

Order: INSECTIVORA   Family: TENRECIDAE

Microgale dobsoni

2n=30

AUTOSOMES: 24 Metacentrics, submetacentrics and subtelocentrics
             4 Acrocentrics

SEX CHROMOSOMES: X Submetacentric
                   Y Small acrocentric

It is not easy to positively identify the X chromosomes by morphology alone, but the Y is the smallest element of the complement.

The karyotypes are gifts of Dr. D. S. Borgaonkar, Johns Hopkins University Medical School, Baltimore, Maryland, USA. The specimens were collected in Madagascar.

A related species, M. talazaci, was found to have the same diploid number as M. dobsoni but missing the long acrocentric pair. Instead, it has a large submetacentric pair, suggestive of reciprocal translocations (Borgaonkar and Gould).

REFERENCES:

1) Borgaonkar, D.S. and Gould, E.: Homozygous reciprocal translocation as a mode of speciation in Microgale Thomas 1883 (Tenrecidae Insectivora). Experientia 24:506, 1968.

2) Borgaonkar, D.S.: Insectivora Cytogenetics. In Comparative Mammalian Cytogenetics (Benirschke, K., ed.), Springer-Verlag, New York, 1969.

Microgale dobsoni

2n=30

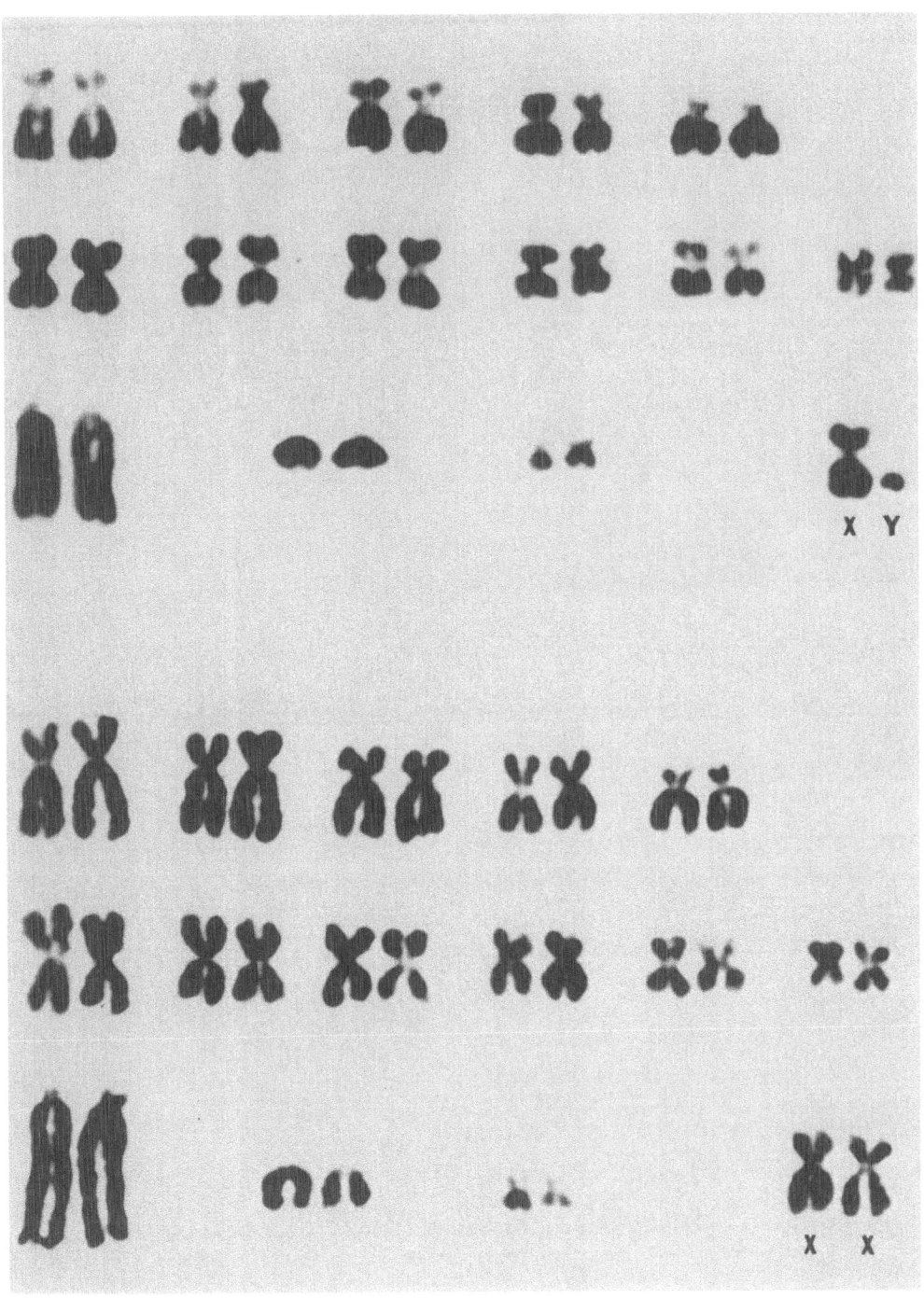

Order: INSECTIVORA

Family: SORICIDAE

Cryptotis parva (Least shrew)
$2n = 52$

Cryptotis parva (Least shrew)

2n=52

AUTOSOMES: 50 Acrocentrics

SEX CHROMOSOMES: X Submetacentric
                   Y Submetacentric

    With the exception of one pair which is extraordinarily long, all the remaining autosomes form a graded series of lengths. The sex chromosomes are easy to identify since they are morphologically unique. The second arm of the Y chromosome is distinctly shorter than the long arm.

    The specimens were donated by Dr. C. H. Conaway, University of Missouri, Columbia, Missouri, USA. They were collected in the vicinity of Columbia, Boone County, Missouri, USA. Lung tissues were used to initiate cell cultures for cytological preparations.

<u>Cryptotis</u> <u>parva</u> (Least shrew)

2n=52

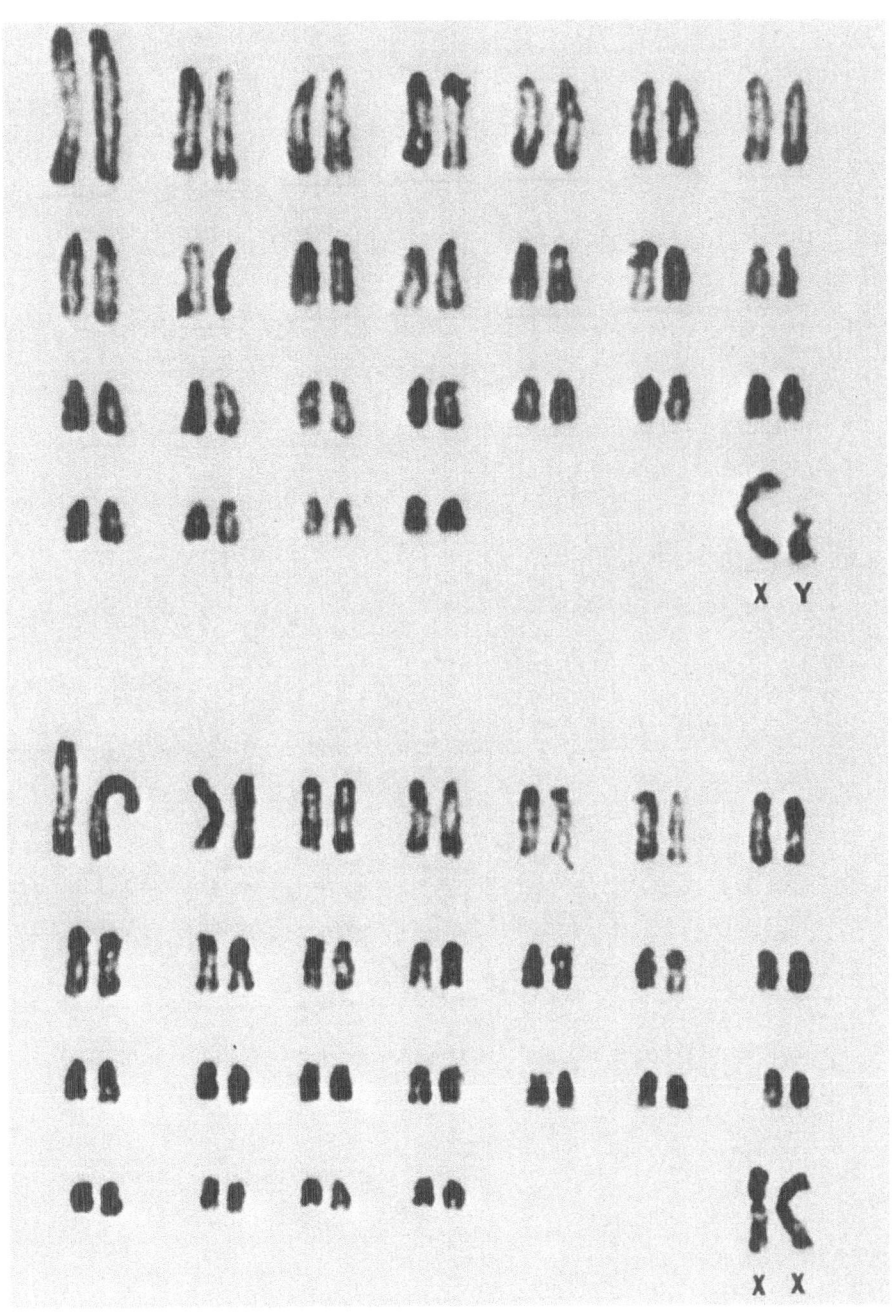

Order: CHIROPTERA

Family: PHYLLOSTOMIDAE

Sturnira lilium (Yellow-shouldered bat)

$2n = 30$

Sturnira lilium (Yellow-shouldered bat)

2n=30

AUTOSOMES: 28 Metacentrics, submetacentrics and subtelocentrics

SEX CHROMOSOMES: X Subtelocentric
                  Y Small subtelocentric

    The autosomes can be classified roughly into three groups: five pairs of large metacentrics and submetacentrics, five pairs of medium or small metacentrics and submetacentrics, and 4 pairs of subtelocentrics. In the last group the smallest pair is more mediocentric than the others. Identification of the sex chromosomes is equivocal.

    The karyotypes are gifts of Dr. Robert J. Baker. The animals were collected from Rancho del Cielo, Tamaulipas, Mexico, and bear voucher specimen numbers 7341 and 7342, Texas Technological University, Lubbock, Texas, USA.

REFERENCES:

1) Baker, R.J.: Karyotypes of bats of the family Phyllostomidae and their taxonomic implications. Southwest. Natural. 12:407, 1967.

2) Hsu, T.C., Baker, R.J. and Utakoji, T.: The multiple sex chromosome system of American leaf-nosed bats (Chiroptera, Phyllostomidae). Cytogenetics 7:27, 1968.

Order: CHIROPTERA   Family: PHYLLOSTOMIDAE

Sturnira lilium (Yellow-shouldered bat)

2n=30

Order: CHIROPTERA

Family: VESPERTILIONIDAE

Eptesicus fuscus (Big brown bat)
2n = 50

Order: CHIROPTERA   Family: VESPERTILIONIDAE

<u>Eptesicus</u> <u>fuscus</u> (Big brown bat)

2n=50

AUTOSOMES: 48 Acrocentrics

SEX CHROMOSOMES: X Submetacentric
                  Y Minute

Three pairs of the autosomes are distinctly smaller than the rest which forms a smooth size gradation. The X chromosomes are the only biarmed elements of the complement, and the Y is the smallest element. Identification of the sex chromosomes is therefore relatively easy.

The karyotypes are gifts of Dr. Robert J. Baker, Texas Technological College, Lubbock, Texas. Bone marrow preparations were used for analysis. Specimens collected in the Houston area showed identical karyotypes from cells in culture.

REFERENCES:

1) Baker, R.J. and Patton, J.L.: Karyotypes and karyotypic variation of North American Vespertilionid bats. J. Mammal. <u>48</u>:270, 1967.

Order: CHIROPTERA   Family: VESPERTILIONIDAE

Eptesicus fuscus (Big brown bat)

2n=50

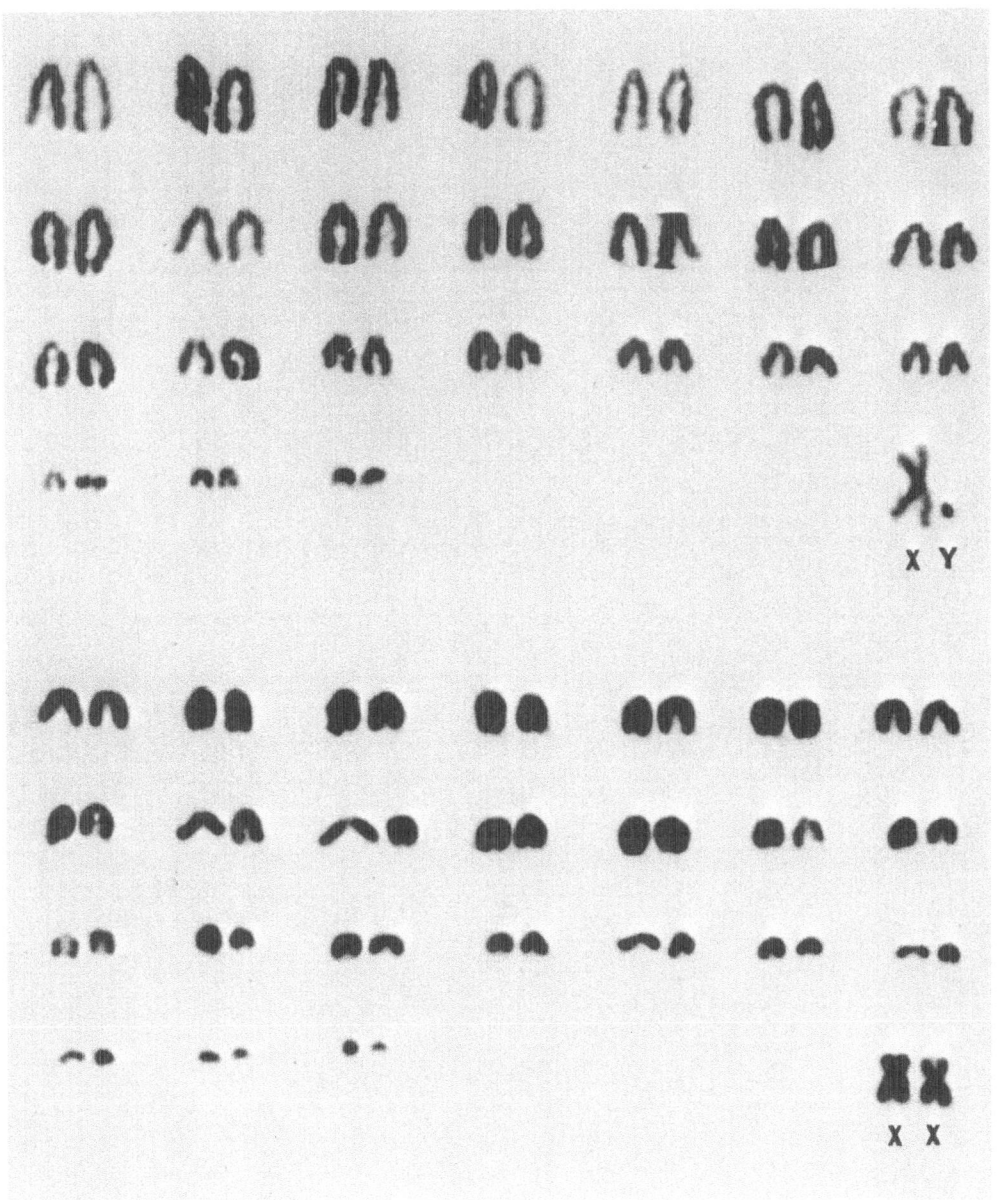

Order: CHIROPTERA

Family: VESPERTILIONIDAE

Lasiurus intermedius (Yellow bat)
$2n = 26$

Order: CHIROPTERA   Family: VESPERTILIONIDAE

<u>Lasiurus</u> <u>intermedius</u> (Yellow bat)

2n=26

AUTOSOMES: 18 Metacentrics and submetacentrics
             6 Acrocentrics

SEX CHROMOSOMES: X Acrocentric
                   Y Minute

     Baker and Patton described that the X chromosomes of <u>L</u>. <u>intermedius</u> is
a submetacentric. Dr. Robert J. Baker, who kindly supplied the karyotypes
of this species presented here, informed us that the original determination
of the X in their paper was erroneous. Upon critical reexamination, the X
chromosomes are found to be the largest acrocentrics of the complement.

     The specimens were collected from 5 miles SE of Brownsville, Cameron
County, Texas, USA, by Dr. Robert J. Baker, bearing voucher specimen numbers
7508♂ and 7504♀. They were stored in Texas Technological University Museum,
Lubbock, Texas, USA.

     When the karyotypes of the species are compared with those of
L. <u>seminolus</u> (Vol. 3, Folio 106), it is obvious that the X chromosomes of the
latter are submetacentric. Other differences are also noted.

REFERENCES:

1) Baker, R.J. and Patton, J.L.: Karyotypes and karyotypic variation of
North American Vespertilionid bats. J. Mammal. <u>48</u>:270, 1967.

Lasiurus intermedius (Yellow bat)

2n=26

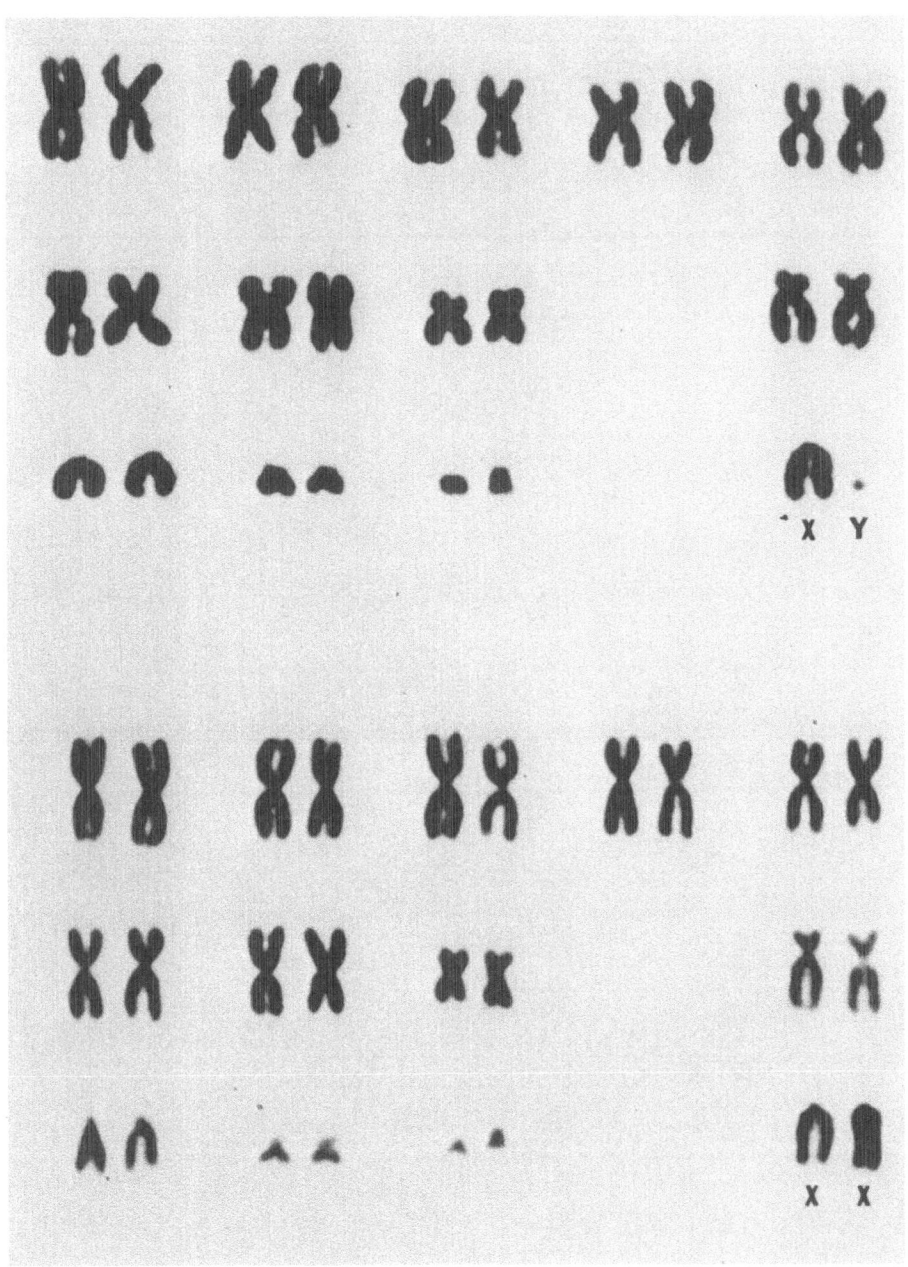

Order: CHIROPTERA

Family: MOLOSSIDAE

Tadarida braziliensis (Mexican free-tailed bat)

$2n = 48$

Order: CHIROPTERA   Family: MOLOSSIDAE

<u>Tadarida braziliensis</u> (Mexican free-tailed bat)

2n=48

AUTOSOMES: 6 Metacentrics and submetacentrics
             40 Acrocentrics and subtelocentrics

SEX CHROMOSOMES: X Metacentric
                  Y Acrocentric

   A number of the "acrocentric" chromosomes have distinct second arm
and one pair bears a secondary constriction near the centromere.
Identification of the X chromosome is not too difficult. Though the X
is morphologically similar to the second pair of largest metacentric
autosomes, it is slightly shorter than these autosomes. The Y chromosome
is the smallest element of the complement.

   The karyotypes presented here are gifts of Dr. Robert J. Baker,
Texas Technological College, Lubbock, Texas, USA. The specimens were
collected in Oklahoma, USA, by Mr. George Rogers.

Tadarida braziliensis (Mexican free-tailed bat)

2n=48

Order: EDENTATA

Family: BRADYPODIAE

Choloepus hoffmanni (Hoffmann's two-toed sloth)

$2n = 49$

Choloepus hoffmanni (Hoffmann's two-toed sloth)

2n=49

AUTOSOMES: 18 Metacentrics and submetacentrics
             30 Acrocentrics

SEX CHROMOSOMES: X Submetacentric
                  Y Translocated onto small metacentric

These karyotypes were kindly donated by Drs. J. Corin and J. Corin-Frederic, University Liège, Belgium. The study was performed on ten animals (fibrous tissue culture) collected in Panama, Costa Rica and Equador. All possessed 49 chromosomes. Females are apparently XO, males have one unpaired element which in size is the 23rd autosome. A Y-23 translocation is inferred; this is supported by meiosis study. Female interphase nuclei are chromatin negative. The X is 5.91%, the Y is 1.58% of the haploid set. The animals were identified by all having six cervical vertebrae.

REFERENCES:

1) Corin-Frederic, J.: L'assortiment chromosomique du paresseux Choloepus hoffmanni Peters. I. Etablissement et étude détaillée du caryotype. II. Mise en évidence d'une formule gonosomique particulière et ses implications dans le mécanisme général de "dosage-compensation" chez les mammifères. Thèse de Doctorat en Sciences Zoologiques, Université de Liège, Belgium, 1968.

2) Corin-Frederic, J.: Technique améliorée pour l'étude des meioses males et en particulier du mode d'association des gonosomes. Genetica 39:345, 1968.

Order: EDENTATA   Family: BRADYPODIDAE

Choloepus hoffmanni (Hoffmann's two-toed sloth)

2n=49

Order: LAGOMORPHA

Family: LEPORIDAE

Lepus americanus (Snowshoe hare)
2n = 48

Lepus americanus (Snowshoe hare)

2n=48

AUTOSOMES: 16 Metacentrics and submetacentrics
             30 Acrocentrics

SEX CHROMOSOMES: X Metacentric
                   Y Acrocentric

Skin cultures of a male (Ravalli County, Montana) and a female (Missoula County, Montana) were kindly made available by Dr. B. W. O'Gara, Missoula, Montana, USA.

In general, the karyotype does not differ appreciably from those of other Lepus species (Folios 6,7); the enumeration of acrocentrics, however, presents difficulties. In other species four acrocentrics only are listed; this depends on one's preference as to how to designate the second group of autosomes here depicted. The results are similar to those of Chang et al.

REFERENCES:

1) Chang, M.C., Pickworth, S. and McGaughey, R.W.: Experimental hybridization and chromosomes of hybrids. In Comparative Mammalian Cytogenetics (Benirschke, K., ed.), Springer-Verlag, N.Y., 1969.

Lepus americanus (Snowshoe hare)

2n=48

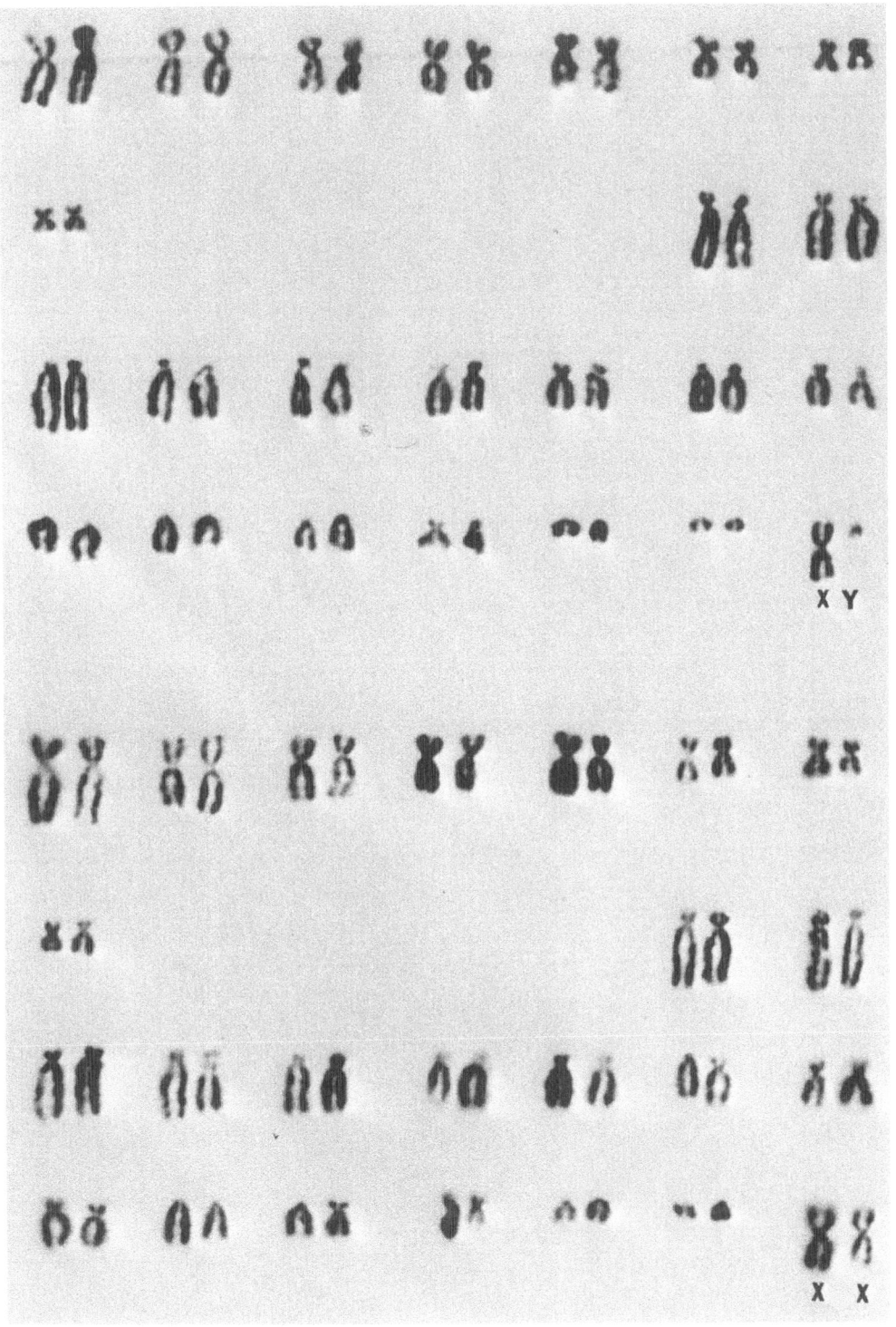

Order: RODENTIA

Family: SCIURIDAE

Sciurus vulgaris (Hokkaido squirrel)

$2n = 40$

Sciurus vulgaris (Hokkaido squirrel)

2n=40

AUTOSOMES: 34 Metacentrics, submetacentrics and subtelocentrics
            4 Acrocentrics

SEX CHROMOSOMES: X Submetacentric
                  Y Acrocentric

A number of chromosome pairs are similar in morphology so that pairing is rather subjective. This also applies to the identification of the X.

The karyotypes presented here are gifts of Dr. M. Sasaki, Hokkaido University, Sapporo, Japan.

REFERENCES:

1) Nadler, C.F. and Sutton, D.A.: Chromosomes of some squirrels (Mammalia-Sciuridae) from the genera Sciurus and Glaucomys. Experientia 23:249, 1967.

2) Sasaki, M., Shimba, H. and Itoh, M.: Notes on the somatic chromosomes of two species of Asiatic squirrels. Chromosome Information Service (Japan) No. 9:6, 1968.

Sciurus vulgaris (Hokkaido squirrel)

2n=40

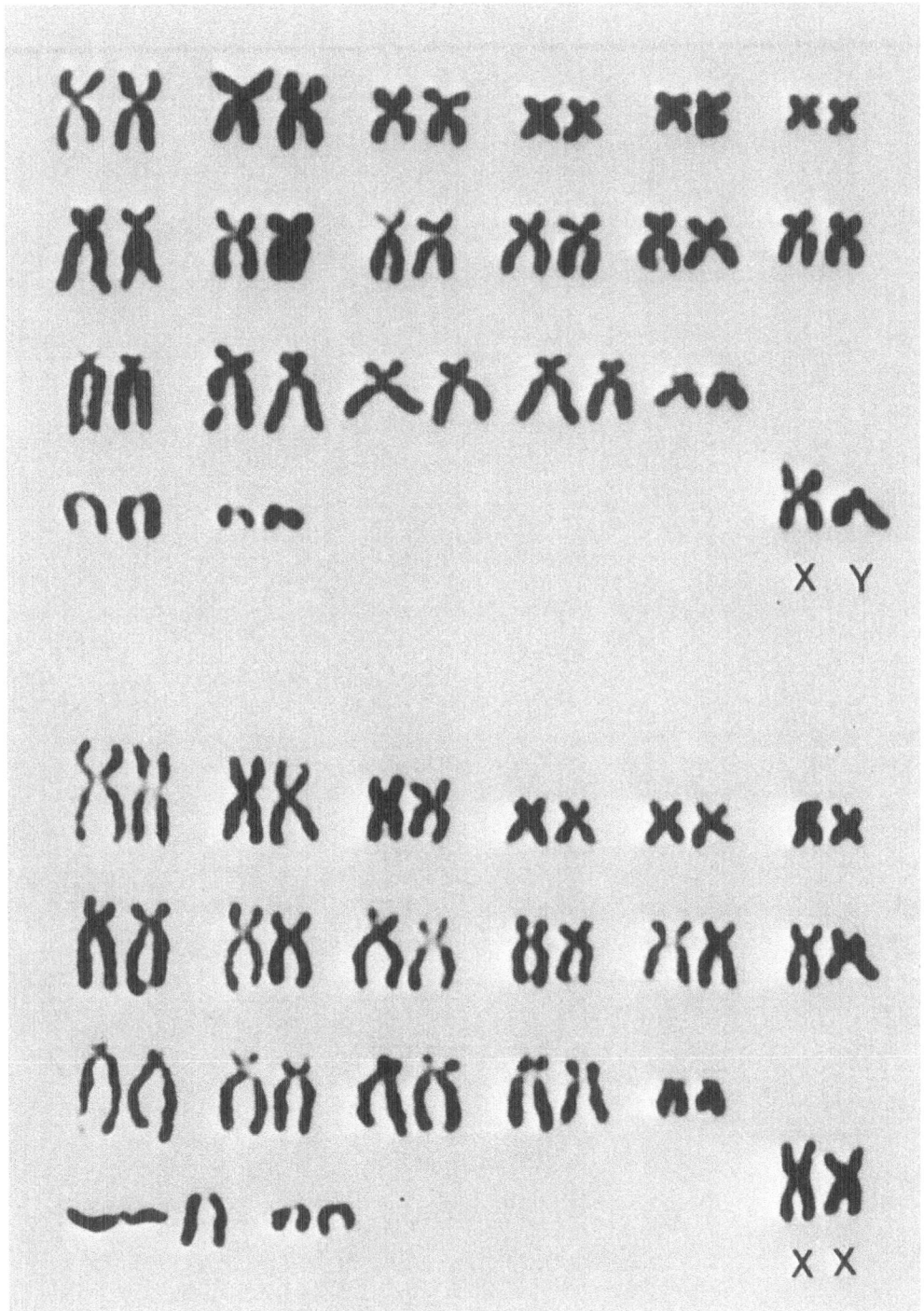

Order: RODENTIA

Family: HETEROMYIDAE

Dipodomys merriami (Merriam's kangaroo rat)
$2n = 52$

Order: RODENTIA   Family: HETEROMYIDAE

Dipodomys merriami (Merriam's kangaroo rat)

2n=52

AUTOSOMES: 50 Metacentrics and submetacentrics

SEX CHROMOSOMES: X Metacentric
                  Y Acrocentric

    Since many chromosomes of this species are similar in morphology,
pairing is very equivocal. However, identification of the Y chromosome
is not difficult.

    The specimens were collected at locality 23 miles West of River Road
and Oatman Road junction, Mohave County, Arizona, USA, in April, 1969.
Bone marrow samples were used for cytological preparations.

Dipodomys merriami (Merriam's kangaroo rat)

2n=52

Order: RODENTIA

Family: HETEROMYIDAE

Perognathus arenarius (Little desert pocket mouse)
$2n = 42$

Perognathus arenarius (Little desert pocket mouse)

2n=42

AUTOSOMES: 24 Metacentrics and submetacentrics
             16 Acrocentrics

SEX CHROMOSOMES: X Submetacentric
                  Y Acrocentric

    Satellite may be observed on some small biarmed chromosomes. Many chromosome pairs are similar in size and in shape, so that identification of individual elements, including the X and the Y, is subjective.

    The karyotypes presented here are gifts of Dr. James L. Patton, Department of Zoology, University of California, Berkeley, California, USA. The specimens (♂, JLP 1569; ♀, JLP 1566) were collected by Dr. Patton at San Felipe, Baja California, Mexico. Bone marrows were used for cytological preparations.

REFERENCES:

1) Patton, J. L.: Karyotypes of five species of pocket mice, genus Perognathus (Rodentia: Heteromyidae), and a summary of chromosome data for the genus. Mammalian Chromosomes Newsletter 11:3, 1970.

Order: RODENTIA   Family: HETEROMYIDAE

<u>Perognathus</u> <u>arenarius</u> (Little desert pocket mouse)

2n=42

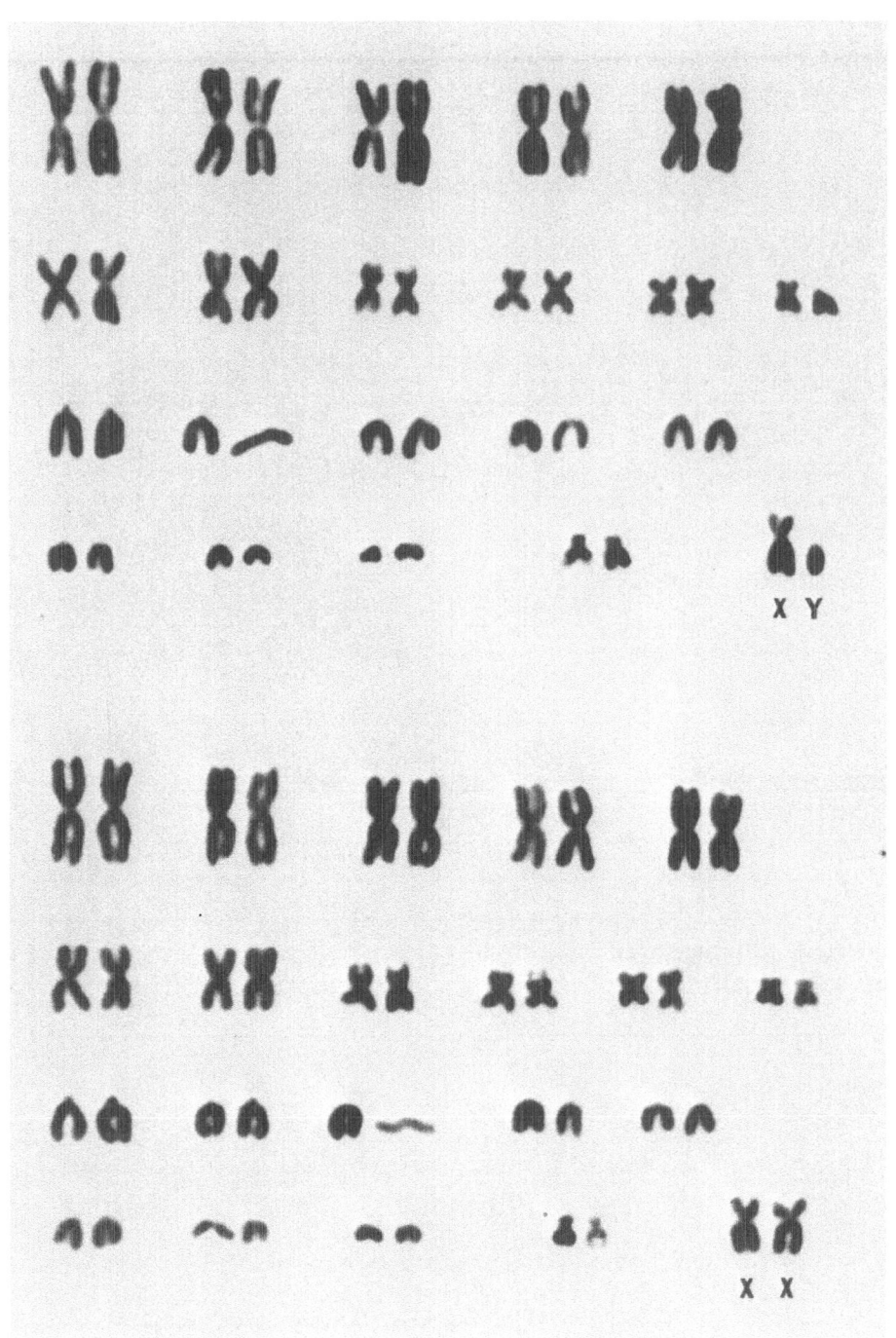

Order: RODENTIA

Family: HETEROMYIDAE

Perognathus spinatus (Spiny pocket mouse)
$2n = 44$

Perognathus spinatus (Spiny pocket mouse)

2n=44

AUTOSOMES: 12 Metacentrics and submetacentrics
             30 Acrocentrics

SEX CHROMOSOMES: X Submetacentric
                  Y Acrocentric

    With one pair which is distinctly shorter, the metacentric autosomes are very similar in shape and in size. A similar situation exists in the acrocentrics: one pair is distinctly smaller than others which form a smooth series in lengths. Identification of the X chromosome is feasible, but the Y chromosome is similar in size to many acrocentric autosomes.

    The karyotypes are gifts of Dr. James L. Patton, Department of Zoology, University of California, Berkeley, California, USA. The specimens (♂, JLP 1554; ♀, JLP 1555) were collected from San Felipe, Baja California, Mexico. Bone marrows were used for cytological preparations.

REFERENCES:

1) Patton, J. L.: Karyotypes of five species of pocket mice, genus Perognathus (Rodentia: Heteromyidae), and a summary of chromosome data for the genus. Mammalian Chromosomes Newsletter 11:3, 1970.

Order: RODENTIA

Family: HETEROMYIDAE

Perognathus spinatus (Spiny pocket mouse)

2n=44

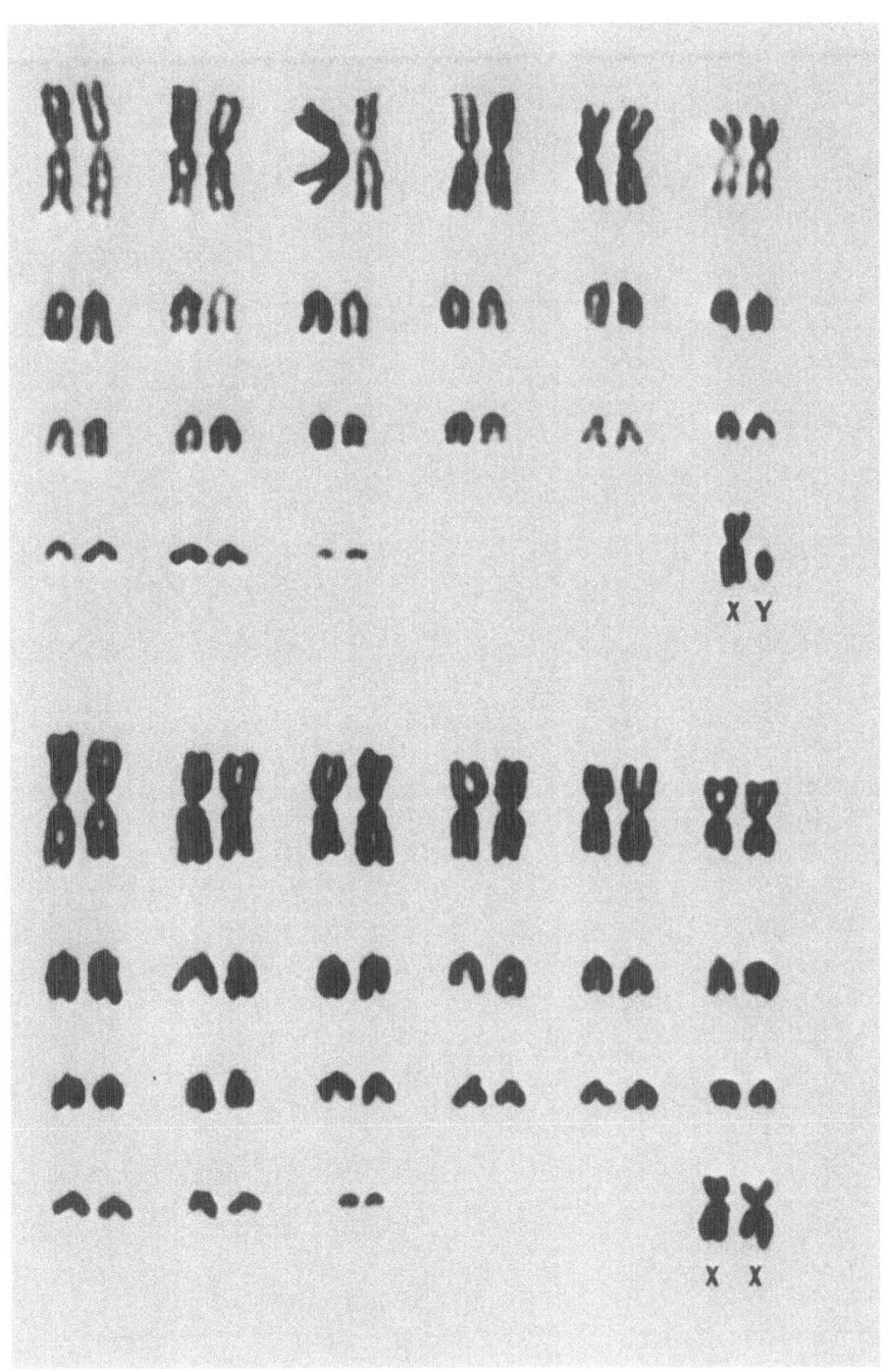

Order: RODENTIA

Family: CRICETIDAE

Cricetus cricetus (European or black-bellied hamster)
$2n = 22$

Cricetus cricetus (European or black-bellied hamster)

2n=22

AUTOSOMES: 18 Metacentrics and submetacentrics
             2 Acrocentrics

SEX CHROMOSOMES: X Submetacentric
                  Y Metacentric

    The karyotypes depicted here were made from bone marrow preparations and were kindly donated by Dr. U. Wolf, University of Freiburg, Germany. The sex chromosomes were identified by radioautography. The X chromosome constitutes about 10-11% of the haploid set, the Y chromosome is about 80% of the size of the X. Five male and five female animals were studied.

REFERENCES:

1) Matthey, R.: Chromosomes des Muridae (Microtinae et Cricetinae). Chromosoma 5:113, 1952.

2) Fredga, K. and Santesson, B.: Male meiosis in the Syrian, Chinese, and European hamsters. Hereditas 52:36, 1964.

3) Wolf, U. and Hepp, D.: DNS-Reduplikationsmuster der somatischen Chromosomen von Cricetus cricetus (L.). Chromosoma 18:438, 1966.

<u>Cricetus</u> <u>cricetus</u> (European or black-bellied hamster)

2n=22

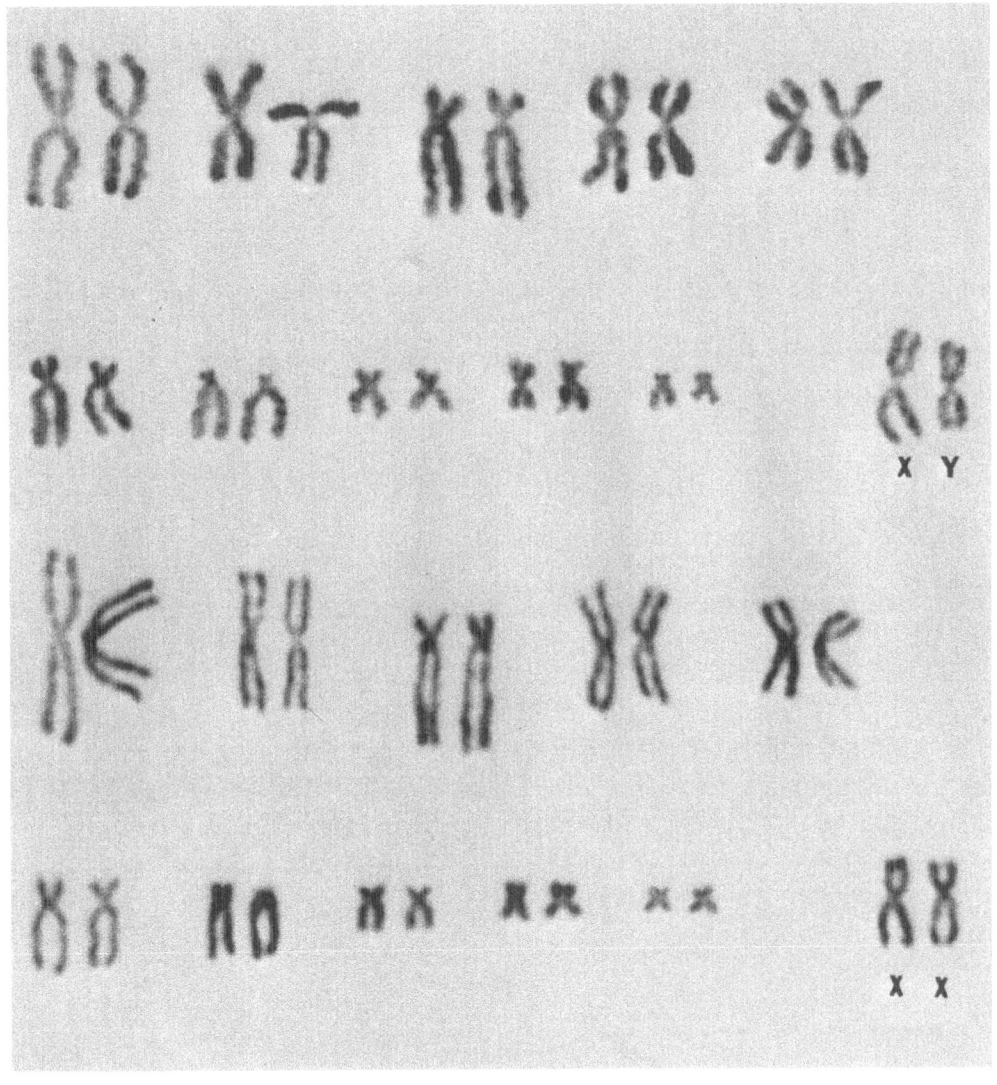

Order: RODENTIA

Family: CRICETIDAE

Neotoma floridana (Eastern wood rat)

$2n = 52$

Order: RODENTIA   Family: CRICETIDAE

<u>Neotoma</u> <u>floridana</u> (Eastern wood rat)

2n=52

AUTOSOMES: 2 Large submetacentrics
                6 Small submetacentrics
               42 Acrocentrics

SEX CHROMOSOMES: X Large submetacentric
                  Y Large metacentric

     A large number of the acrocentric chromosomes possess knob-like second
arm. It is not easy to positively identify the X chromosome because it may
be confused with the large submetacentric autosomes. However, the Y
chromosome is unique in its morphology and can be identified unequivocally.

     The specimens were collected from Payne County, Oklahoma, USA, by
Mr. R. E. Martin. Lung tissues were used to initiate cell cultures for
cytological preparations. Specimens collected at College Station, Texas, USA,
and from Tampa, Florida, USA, showed identical karyotypes.

REFERENCES:

1) Matthey, R.: Les chromosomes des Muridae. Rev. suisse Zool. <u>60</u>:225, 1953.

2) Baker, R.J. and Mascarello, J.T.: Karyotypic analyses of the genus
<u>Neotoma</u> (Cricetidae, Rodentia). Cytogenetics <u>8</u>:187, 1969.

<u>Neotoma</u> <u>floridana</u> (Eastern wood rat)

2n=52

Order: RODENTIA

Family: CRICETIDAE

Peromyscus truei (Pinon mouse)
2n = 48

Peromyscus truei (Piñon mouse)

2n=48

AUTOSOMES: 16 Submetacentrics
             30 Acrocentrics

SEX CHROMOSOMES: X Submetacentric
                   Y Acrocentric

The karyotypes presented here resemble those of P. gossypinus
(Folio No. 66). They were from lung cultures of animals collected at
Monte Rio, Sonoma County, California, USA, by Dr. Murray L. Johnson in 1967.

Samples from several localities of California, Utah, Oklahoma, and
Northern New Mexico showed identical karyotypes. However, specimens from
Southwestern New Mexico (identified as P. t. truei) showed a karyotype
similar to that of P. californicus (Folio No. 115). The distribution range
of the latter is not known, since only one sample is available from
Silver City, New Mexico, USA.

REFERENCES:

1) Hsu, T.C. and Arrighi, F.E.: Chromosomes of Peromyscus (Rodentia,
Cricetidae) I. Evolutionary trends in twenty species. Cytogenetics 7:
417, 1968.

Order: RODENTIA

Family: CRICETIDAE

Peromyscus truei (Piñon mouse)

2n=48

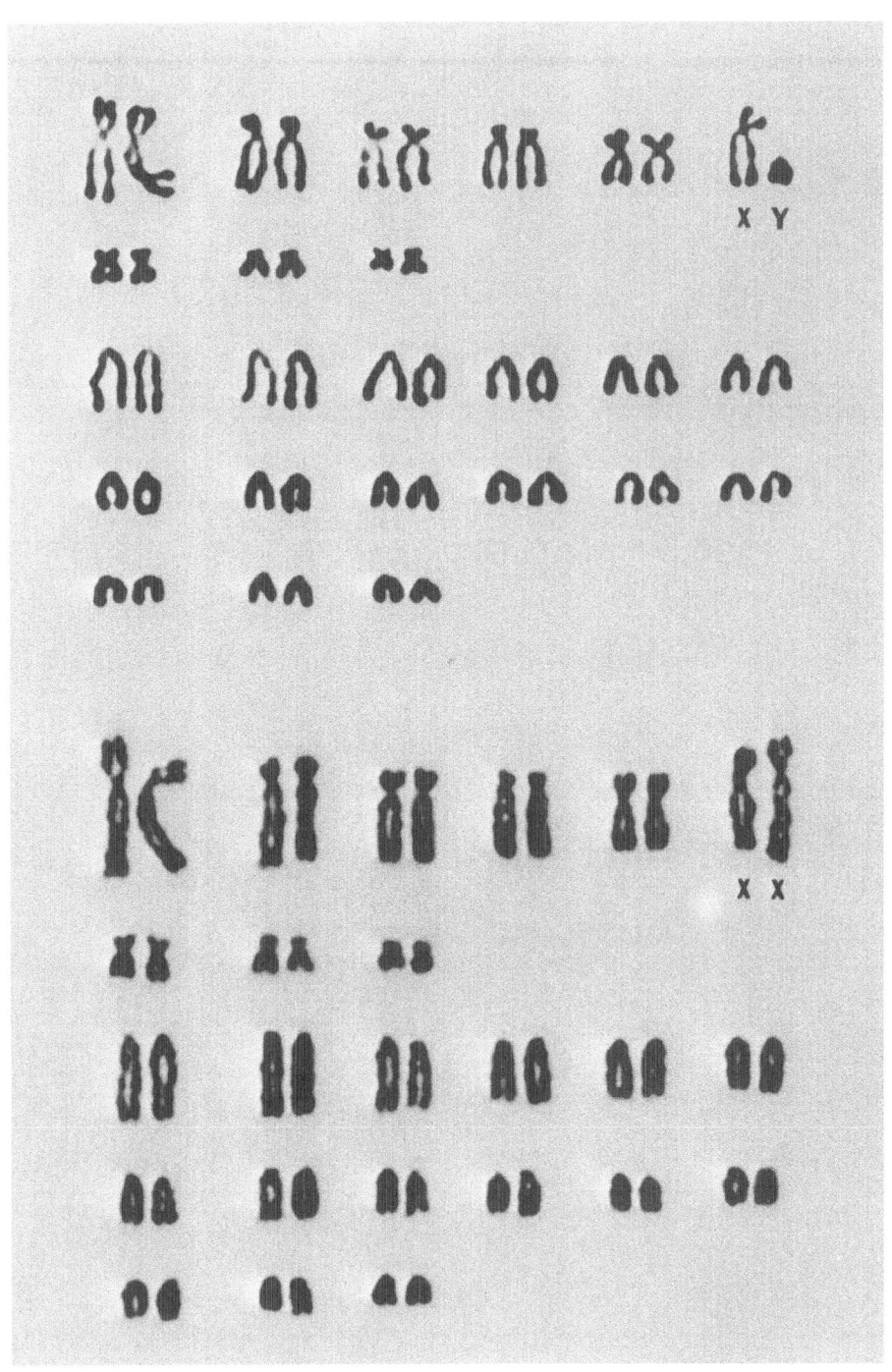

Order: RODENTIA

Family: CRICETIDAE

Reithrodontomys megalotis (Western harvest mouse)
$2n = 42$

Volume 4, Folio 169, 1970

Reithrodontomys megalotis (Western harvest mouse)

2n=42

AUTOSOMES: 40 Metacentrics and submetacentrics

SEX CHROMOSOMES: X Large submetacentric
                  Y Subtelocentric

With two exceptional pairs whose arms are highly unequal, all other autosomes are more or less equal-armed. Secondary constriction can be found on some autosomes near the centromeres. Identification of the sex chromosomes is unequivocal. In female cells, the two X chromosomes usually display a difference in morphology.

Matthey determined the diploid number of R. megalotis as 44. This number was confirmed by Blanks and by Shellheimer. However, in populations from California, USA, Blanks and Shellheimer found that the somatic chromosome number varied depending on the number of minute chromosomes present. The lowest number was 42 (no minutes), but individuals were found to possess 43 (1 minute), 44 (2 minutes), 45 (3 minutes), and 46 (4 minutes). In view of the variability, it is probably best to regard the diploid number as 42 instead of 44. Blanks and Shellheimer voiced the same opinion.

In populations of New Mexico and Arizona, similar supernumerary minutes were found. The karyotypes presented here were from specimens collected in the vicinity of Silver City, New Mexico, USA. Lung cultures were initiated for cytological preparations.

REFERENCES:

1) Blanks, G.A.: A cytotaxonomical and morphological study of the harvest mice of the lower Salinas Valley: Reithrodontomys megalotis longicaudus and R. m. distichlis. M.A. Thesis, San Jose State College, 1967.

2) Shellheimer, H.S.: Cytotaxonomic studies of the harvest mice of the San Francisco Bay region. J. Mammal. 48:549, 1967.

3) Blanks, G.A. and Shellheimer, H.S.: Chromosome polymorphism in California populations of harvest mice. J. Mammal. 49:726, 1968.

Order: RODENTIA   Family: CRICETIDAE

Reithrodontomys megalotis (Western harvest mouse)

2n=42

Order: RODENTIA

Family: CRICETIDAE

Psammomys obesus (Sand rat)
$2n = 48$

<u>Psammomys obesus</u> (Sand rat)

2n=48

AUTOSOMES: 28 Metacentrics and submetacentrics
             18 Acrocentrics

SEX CHROMOSOMES: X Large submetacentric
                   Y Small metacentric

    The X chromosome is the largest of the entire complement, so that it
is very easy to identify. The Y chromosome is similar in size to the
smallest biarmed autosomes, but its centromeric position is more central.

    The animals were donated by Dr. Knut Schmidt-Nielsen, Duke University,
North Carolina, USA. These were bred in captivity. Lung cultures were
initiated for karyological studies.

REFERENCES:

1) Smith, A.G., Hackel, D.B. and Schmidt-Nielsen, K.: Chromosomes of the
sand rat (<u>Psammomys obesus</u>). Canad. J. Genet. & Cytol. <u>8</u>:756, 1966.

Psammomys obesus (Sand rat)

2n=48

Order: RODENTIA

Family: CRICETIDAE

Clethrionomys gapperi (Gapper's red-backed mouse)
$2n = 56$

<u>Clethrionomys</u> <u>gapperi</u> (Gapper's red-backed mouse)

2n=56

AUTOSOMES: 2 Small submetacentrics
             52 Acrocentrics

SEX CHROMOSOMES: X Acrocentric
                  Y Acrocentric

     The X chromosomes are among the largest elements of the complement, and
the Y, smallest. Other than the single pair of small submetacentrics,
pairing of the autosomes is arbitrary. The karyotypes of <u>C</u>. <u>gapperi</u> are
indistinguishable from those of <u>C</u>. <u>rufocanus</u> <u>bedfordi</u> of the Orient
(Vol. 3, Folio No. 119). Recent studies on the chromosomes of another
species, <u>C</u>. <u>occidentalis</u>, revealed again identical karyotypes. The
karyotypes of <u>C</u>. <u>occidentalis</u> will, therefore, not be presented in the
future.

     The male specimen was collected in Clallam County, Washington, USA, and
the female, from East Pine, British Columbia, Canada, both by Dr. Murray L.
Johnson. Lung tissues were used to initiate cell cultures for cytological
preparations. A female specimen collected from Boulder County, Colorado,
USA, showed identical karyotypes.

REFERENCES:

1) Matthey, R.: les chromosomes des Muridae. Révision critique et
matériaux nouveaux pour servir à l'histoire de l'évolution chromosomique
chez ces Rongeurs. Rev. Suisse Zool. <u>60</u>:466, 1953.

Clethrionomys gapperi (Gapper's red-backed mouse)

2n=56

Order: RODENTIA

Family: CRICETIDAE

Clethrionomys (Evotomys) glareolus (Bank vole)
2n = 56

Clethrionomys (Evotomys) glareolus (Bank vole)

2n=56

AUTOSOMES: 2 Metacentrics
              52 Acrocentrics

SEX CHROMOSOMES: X Acrocentric
                   Y Metacentric

Karyotypes were prepared from fibroblast cultures of lung tissue of more than 10 animals. All gave similar chromosomal features which were identical to those from bone marrow preparations. The karyotypes were kindly supplied by Dr. U. Wolf, University of Freiburg, Germany. The X chromosomes have been positively identified by radioautography.

REFERENCES:

1) Matthey, R.: Cytologie comparée systématique et phylogenie des microtinae (Rodentia-Muridae). Rev. Suisse Zool. 64:39, 1957.

2) Matthey, R. and Renaud, P.: Le type de digamétie mâle et les chromosomes chez deux Campagnols. C. R. Soc. Biol. 120:595, 1935.

3) Renaud, P.: La formule chromosomiale chez sept espèces de Muscardinidae et de Microtinae indigènes. Rev. Suisse Zool. 45:349, 1938.

4) Schmid, W. and Leppert, M.F.: Karyotyp, Heterochromatin und DNS=Werte bei 13 Arten von Wühlmäusen (Microtinae, Mammalia-Rodentia). Arch. Julius Klaus-Stiftung 43:88, 1968.

<u>Clethrionomys</u> (<u>Evotomys</u>) <u>glareolus</u> (Bank vole)

2n=56

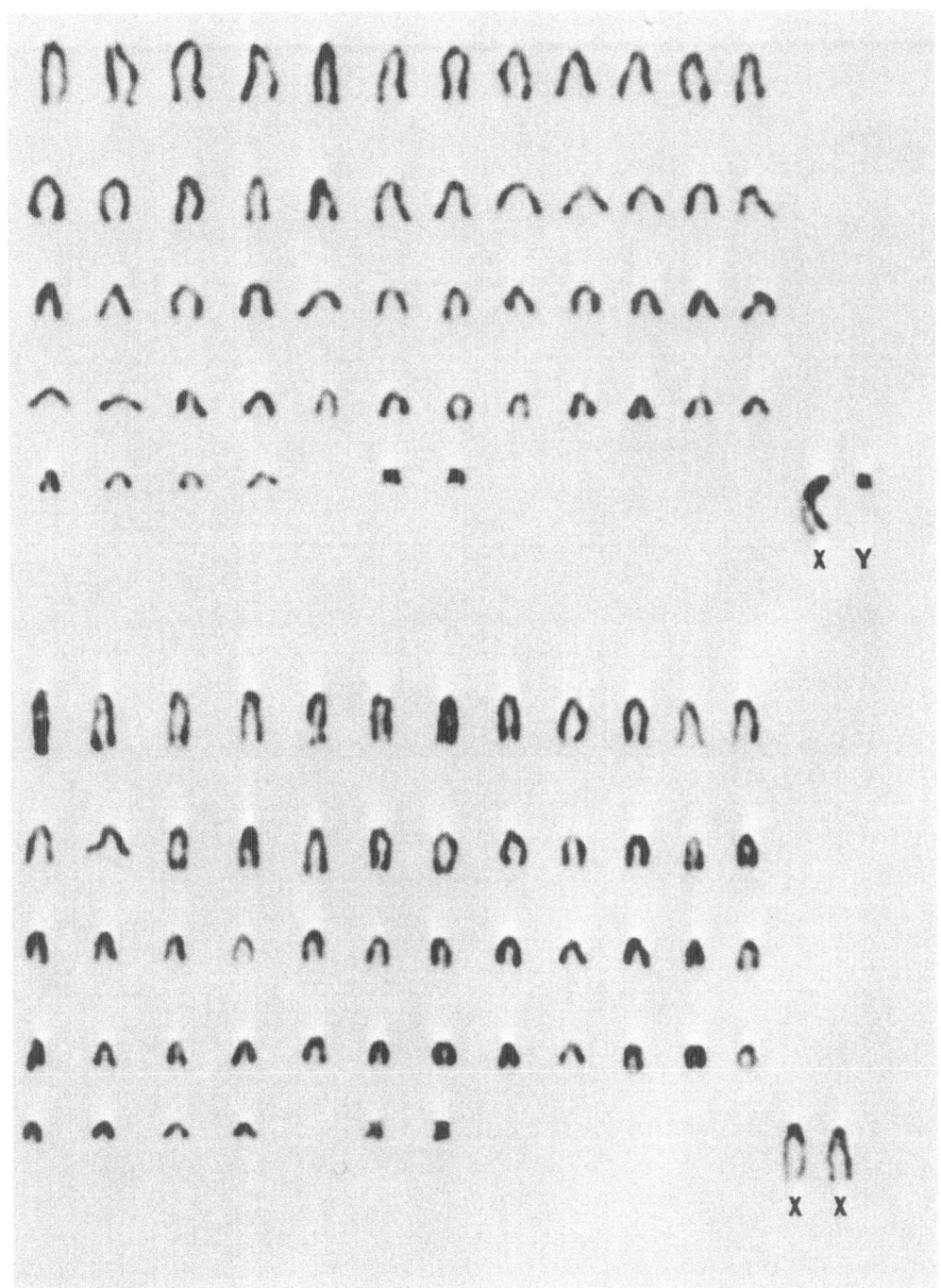

Order: RODENTIA

Family: CRICETIDAE

Microtus arvalis (Common vole)

$2n = 46$

Volume 4, Folio 173, 1970

<u>Microtus arvalis</u> (Common vole)

2n=46

AUTOSOMES: 36 Metacentrics and submetacentrics
             8 Acrocentrics

SEX CHROMOSOMES: X Metacentric
                  Y Minute (? acrocentric)

These karyotypes were kindly supplied by Dr. U. Wolf, University of Freiburg, Germany, and were made from animals caught in that vicinity. A different type of sex determining mechanism ($X_1X_2Y_1Y_2/X_1X_1X_2X_2$) has been suggested by Raicu <u>et</u> <u>al</u>.

REFERENCES:

1) Renaud, P.: La formule chromosomiale chez sept espèces de Muscardinidae et de Microtinae indigènes. Rev. Suisse Zool. <u>45</u>:349, 1938.

2) Raicu, P., Kirillova, M. and Hamar, M.: A new chromosomal sex-determining mechanism in <u>Microtus arvalis</u> Pallas. Genetica <u>40</u>:97, 1969.

3) Schmid, W. and Leppert, M.F.: Karyotyp, Heterochromatin und DNS=Werte bei 13 Arten von Wühlmäusen (<u>Microtinae</u>, <u>Mammalia-Rodentia</u>). Arch Julius Klaus-Stiftung <u>43</u>:88, 1968.

Microtus arvalis (Common vole)

2n=46

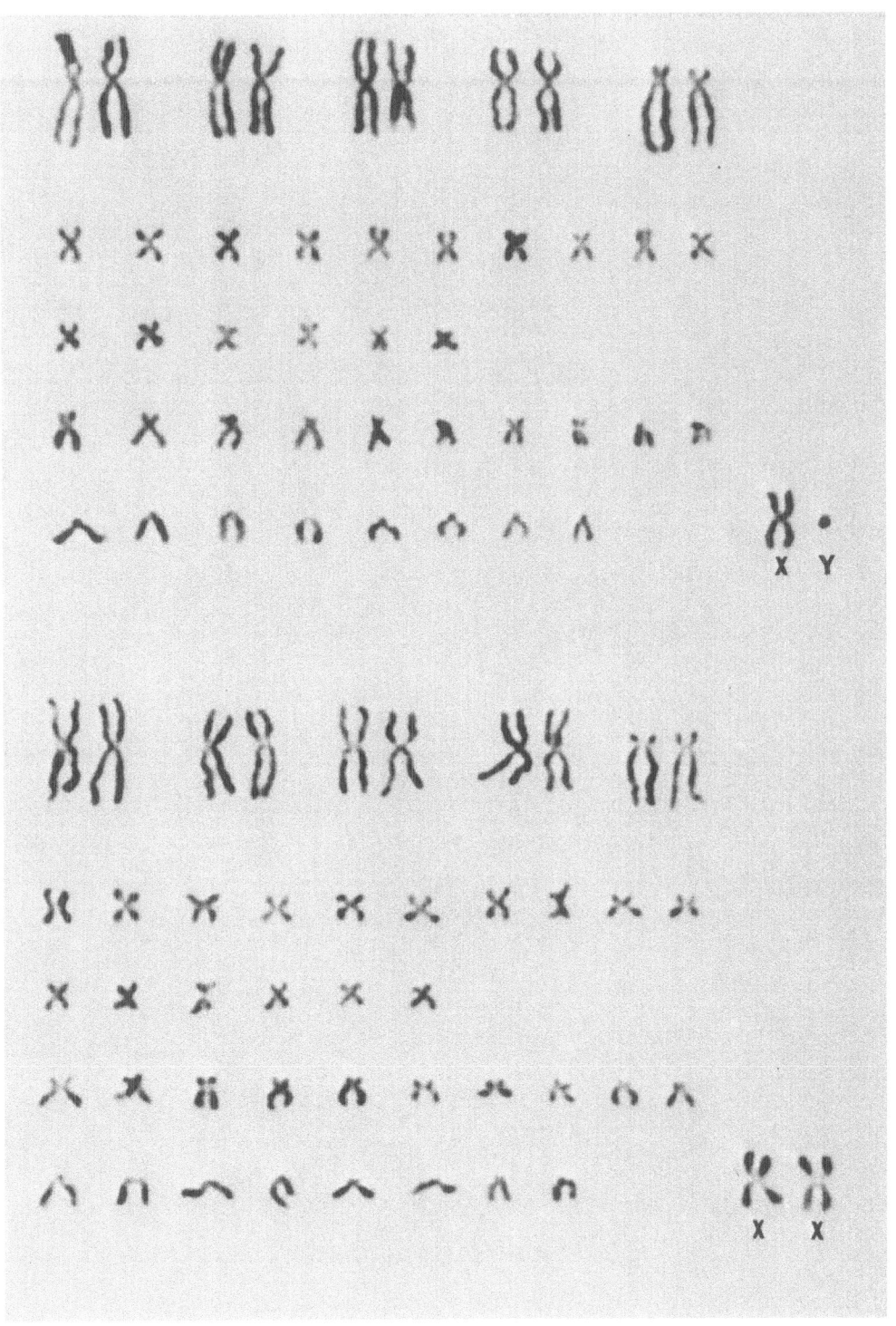

Order: RODENTIA

Family: CRICETIDAE

Microtus montebelli (Korean vole)
2n = 30

Order: RODENTIA   Family: CRICETIDAE

<u>Microtus</u> <u>montebelli</u> (Korean vole)

2n=30

AUTOSOMES: 28 Metacentrics, submetacentrics and subtelocentrics

SEX CHROMOSOMES: X Metacentric
                  Y Acrocentric

One pair of autosomes is considerably longer than the rest. Several other pairs of autosomes can be identified by their morphology; but pairing of most chromosomes is rather subjective. Identification of the X is also equivocal.

The karyotypes presented here are gifts of Dr. Tadashi Utakoji, Cancer Institute, Tokyo, Japan. The specimens were collected in Japan.

Microtus montebelli (Korean vole)

2n=30

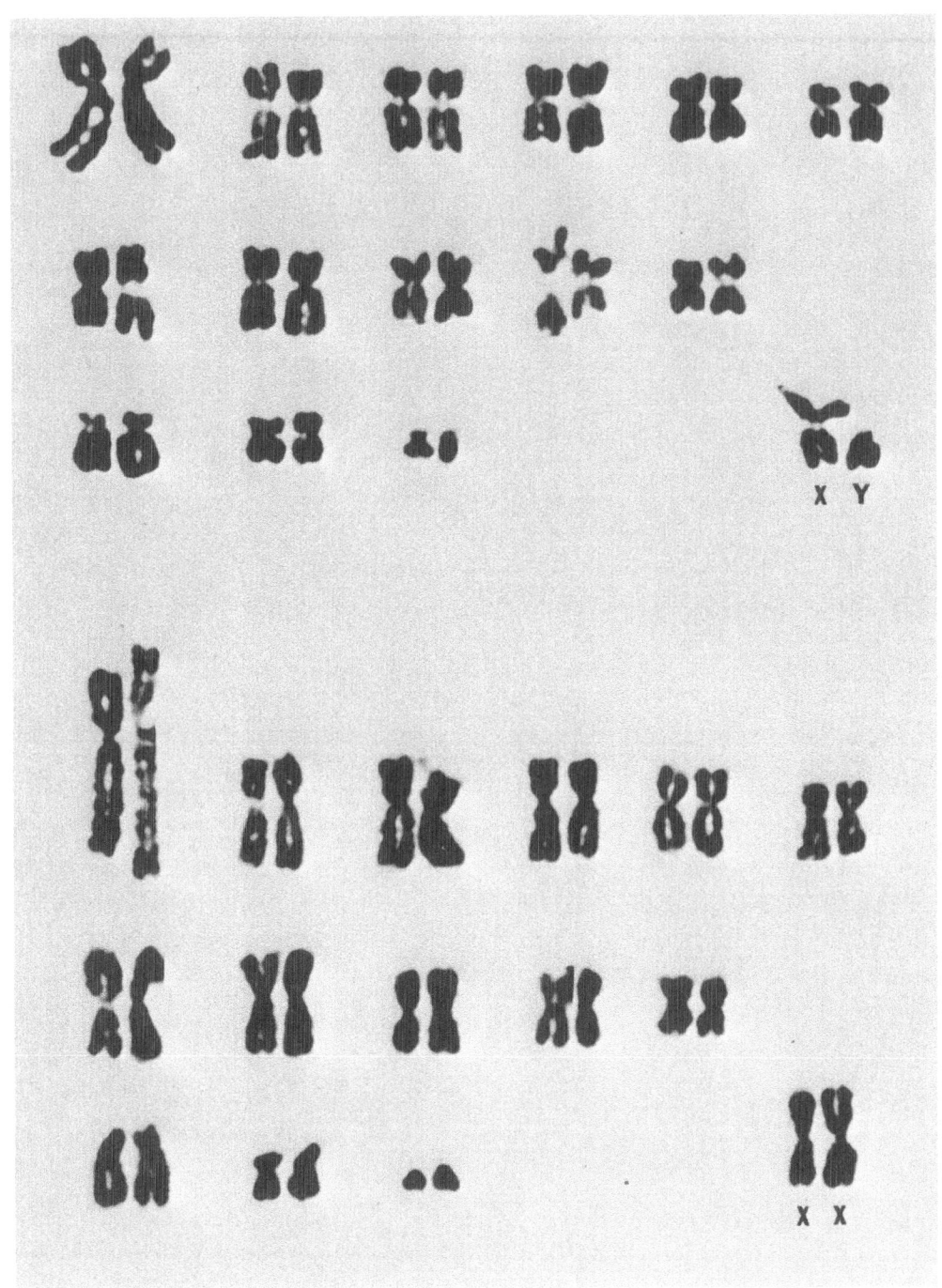

Order: RODENTIA

Family: CRICETIDAE

Microtus oeconomus (Northern vole)

$2n = 30$

Microtus oeconomus (Northern vole)

2n=30

AUTOSOMES: 24 Metacentrics and submetacentrics
             4 Acrocentrics

SEX CHROMOSOMES: X Metacentric
                   Y Acrocentric

Two males and three females from central Germany were karyotyped from cells of embryonal cultures. Lung and bone marrow gave similar results. The X chromosomes were identified by radioautography. Constant secondary constrictions are seen in the long arms of the largest and smallest autosomes. These karyotypes have been supplied by Dr. U. Wolf, University of Freiburg, Germany. The possibility has been suggested that M. oeconomus is identical to M. ratticeps (root vole).

REFERENCES:

1) Makino, S.: Studies on murine chromosomes, VI. Morphology of the sex chromosomes in two species of Microtus. Annotat. Zool. Japan 23:63, 1950.

2) Matthey, R.: Chromosomes des Muridae (Microtinae et Cricetinae). Chromosoma 5:113, 1952.

3) Rausch, R.L. and Rausch, V.R.: On the biology and systematic position of Microtus abbreviatus Miller, a vole endemic to the St. Matthew Islands, Bering Sea. Z. Säugetierk. 33:65, 1968.

4) Brink, F.H. v.d.: A Field Guide to the Mammals of Britain and Europe. Collins, London, 1967.

5) Schmid, W. and Leppert, M.F.: Karyotyp, Heterochromatin und DNA=Werte bei 13 Arten von Wühlmäusen (Microtinae, Mammalia-Rodentia). Arch. Julius Klaus-Stiftung 43:88, 1968.

Order: RODENTIA   Family: CRICETIDAE

Microtus oeconomus (Northern vole)

2n=30

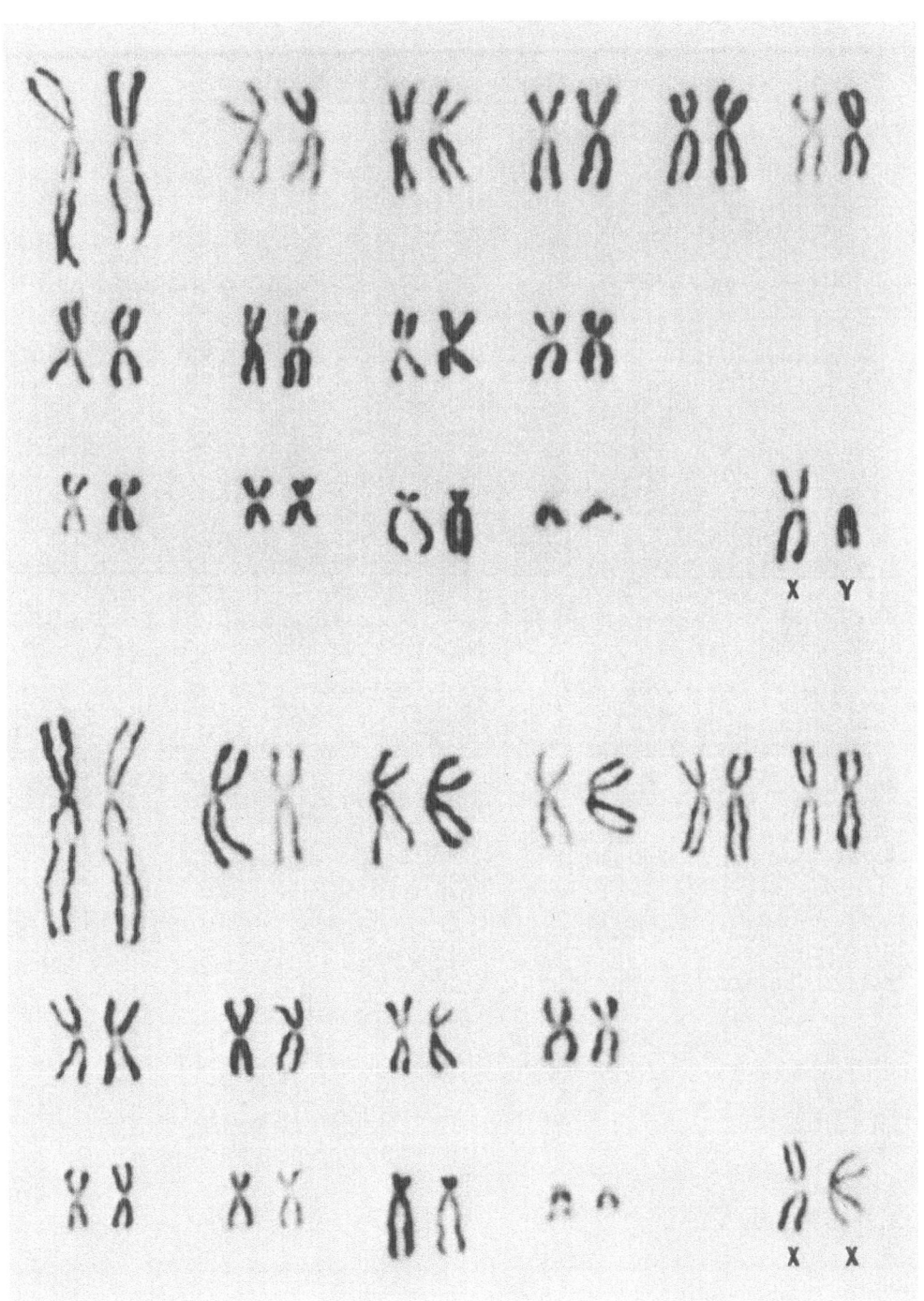

Order: RODENTIA

Family: MURIDAE

Apodemus sylvaticus (Field mouse)

2n = 48

Apodemus sylvaticus(Field mouse)

2n=48

AUTOSOMES: 46 Acrocentrics

SEX CHROMOSOMES: X Acrocentric
                  Y Acrocentric

     The X chromosome is the longest element of the complement, much longer
than the longest autosome. Therefore even though the karyotypes of this
species consist of all acrocentric chromosomes, identification of the X
poses no problem. Identification of the Y, however, is subjective.

     The karyotypes presented here are gifts of Dr. N. P. Bishun, Department
of Haematology, Institute of Child Health, London, England.

REFERENCES:

1) Matthey, R.: La formule chromosomiale et les hétérochromosomes chez
les Apodemus européens. Z. Zellf. mikr. Anat. 25:501, 1936.

2) Koller, P.: The genetical and mechanical properties of sex chromosomes.
VII. Apodemus sylvaticus and A. hebridensis. J. Genet. 41:375, 1941.

Apodemus sylvaticus(Field mouse)

2n=48

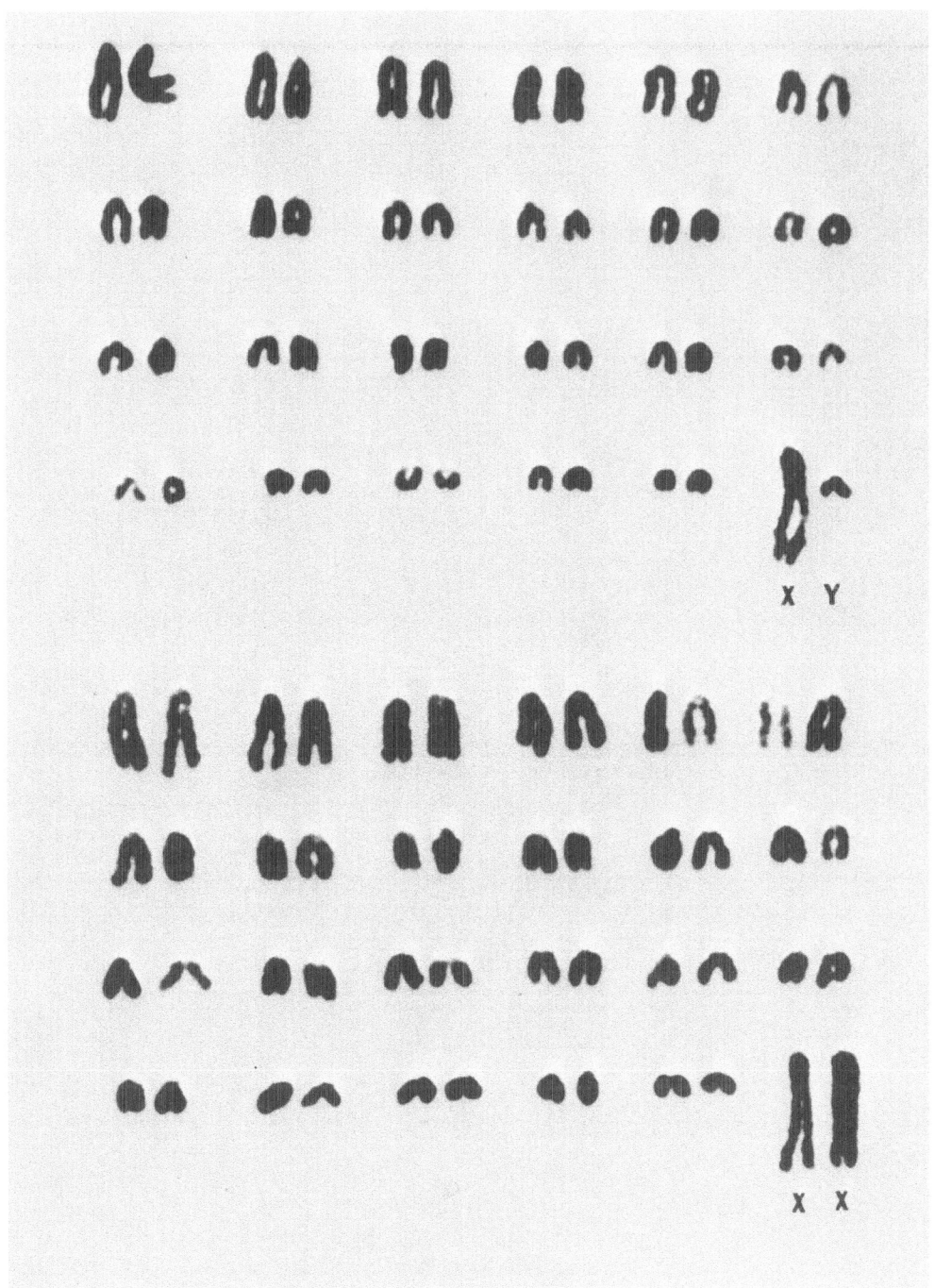

Order: RODENTIA

Family: MURIDAE

Mus poschiavinus (Tobacco mouse)
$2n = 26$

Order: RODENTIA   Family: MURIDAE

<u>Mus poschiavinus</u> (Tobacco mouse)
2n=26

AUTOSOMES: 14 Metacentrics and submetacentrics
             10 Acrocentrics

SEX CHROMOSOMES: X Acrocentric
                   Y Acrocentric

     This species is known only to the Val Poschiavo (Puschlav) in the
Grisons/Switzerland. The animals were originally trapped by Professor
E.von Lehmann,Zoologisches Forschungsinstitut und Museum Alexander Koenig,
Bonn,Germany. The karyotype of the male is from a spermatogonial metaphase,
that of the female from a squash preparation of the spleen. Both are gifts
of Professor Alfred Gropp and Dr.U.Tettenborn,Bonn,Germany.

     It is of interest to compare the karyotypes of this species with its
relative,<u>Mus musculus</u> (Folio 17) . Obviously Robertsonian processes have
occurred at least seven times. The two species can be hybridized with
semisterility of the F_1 offspring.

REFERENCES:

1) Gropp,A.: Cytologic mechanism of karyotype evolution in insectivores,
additional presentation of a case of Robertsonian variation in mice.
In "<u>Comparative Mammalian Cytogenetics</u>" (Benirschke,K.,ed.), Springer-
Verlag, New York, 1969.
2) Gropp,A., Tettenborn,U. and von Lehmann,E.: Chromosomenvariation
vom Robertson'schen Typus bei der Tabakmaus,<u>M.poschiavinus</u>, und ihren
Hybriden mit der Laboratoriumsmaus. Cytogenetics <u>9</u>: 9, 1970.

<u>Mus</u> <u>poschiavinus</u> (Tobacco mouse)
2n=26

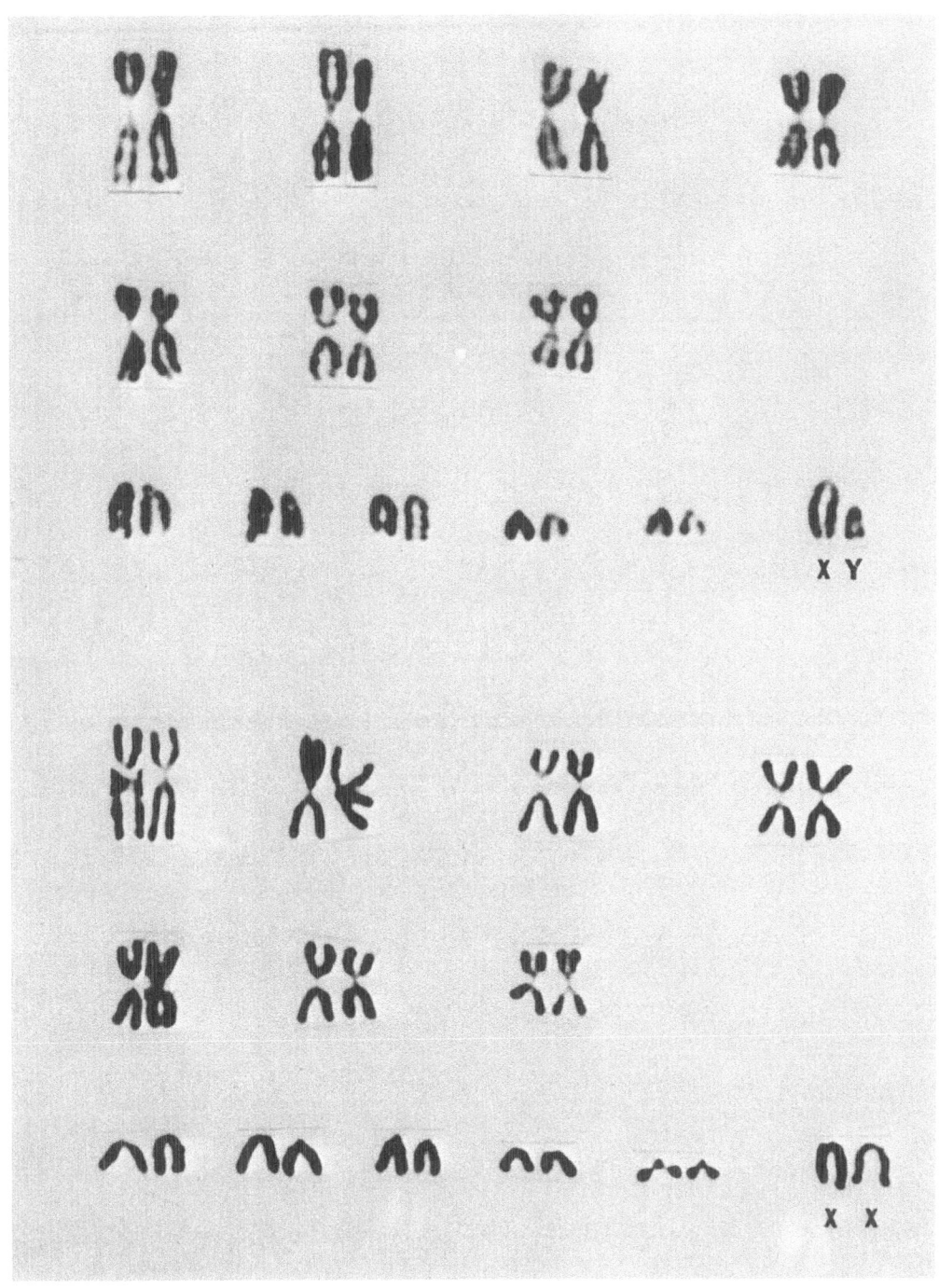

Order: CARNIVORA

Family: CANIDAE

Atelocynus microtis (Round-eared dog, small-eared dog)
2n = 74

Order: CARNIVORA   Family: CANIDAE

Atelocynus microtis (Round-eared dog, small-eared dog)

2n=74

AUTOSOMES: 72 Acrocentrics

SEX CHROMOSOMES: X Submetacentric
                  Y Submetacentric

    Skin biopsies of one male and one female specimen were kindly made
available by Dr. C. Gray, National Zoological Park, Washington, D.C., USA.
The male had consistently 74 chromosomes in this skin culture; a few of
the female cells possessed two additional minute chromosomes. Because of
the difficulties encountered in the culture of this animal we believe these
few cells to be abnormal. Perhaps they represent a phenomenon similar to
that described in the red fox. In several of the smaller autosomes distinct
short arms are seen. There are no marker chromosomes.

REFERENCES:

1) Wurster, D.H.: Cytogenetic and phylogenetic studies in carnivora.
In Comparative Mammalian Cytogenetics (Benirschke, K., ed.), Springer-Verlag,
N. Y., 1969.

2) Wurster, D.H. and Benirschke, K.: Comparative cytogenetic study in the
Order Carnivora. Chromosoma 24:336, 1968.

Atelocynus microtis (Round-eared dog, small-eared dog)

2n=74

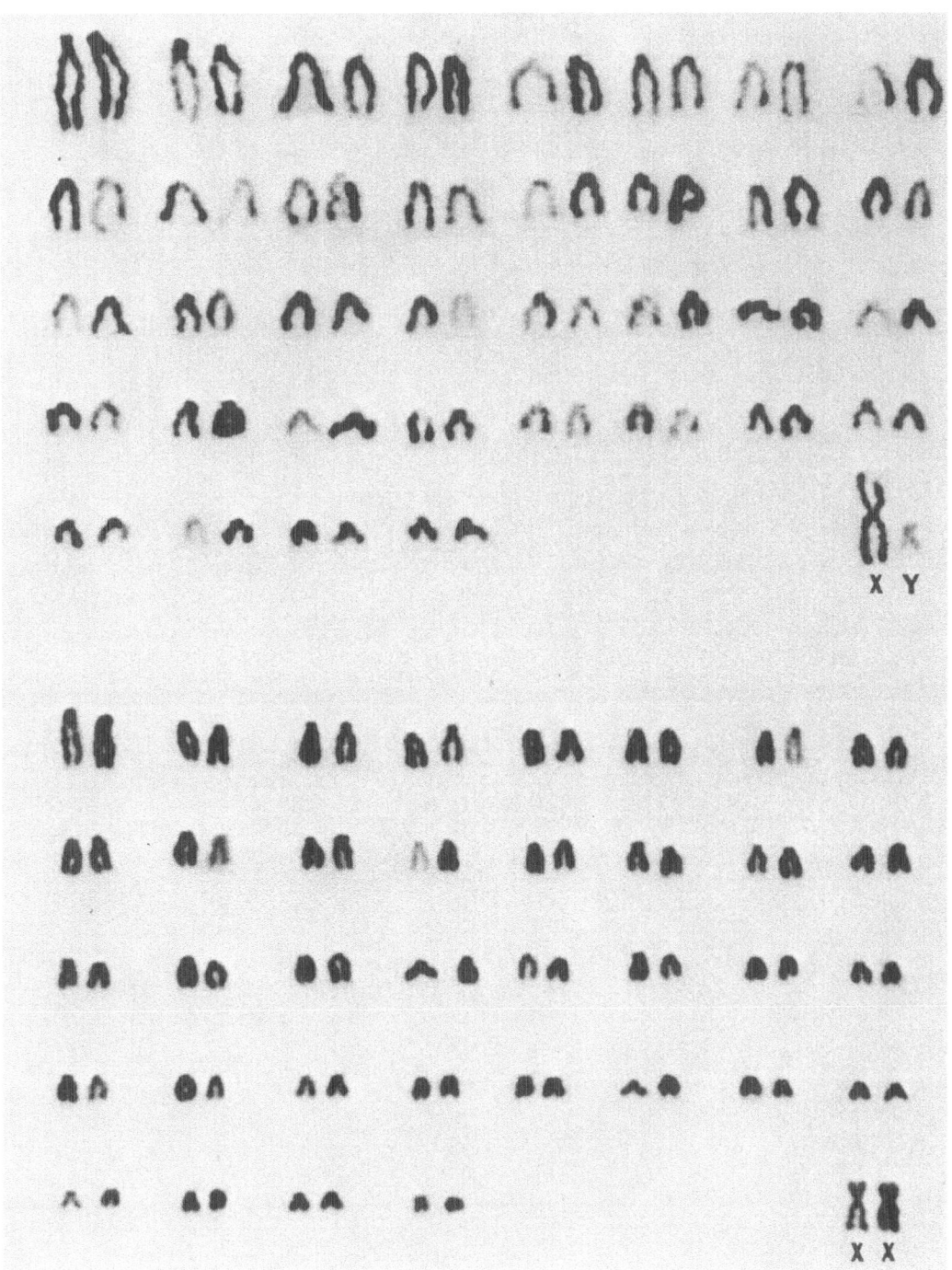

Order: CARNIVORA

Family: CANIDAE

Lycaon pictus (African or Cape hunting dog)
$2n = 78$

Order: CARNIVORA   Family: CANIDAE

Lycaon pictus (African or Cape hunting dog)

2n=78

AUTOSOMES: 76 Acrocentrics

SEX CHROMOSOMES: X Metacentric
                  Y Metacentric

    Skin biopsies from one male and one female hunting dog were kindly
provided by Dr. C. Gray, National Zoological Park, Washington, D.C., USA.
It is difficult to determine whether all small autosomes are indeed
acrocentric; the last three pairs may have short arms.

REFERENCES:

1) Wurster, D.H.: Cytogenetic and phylogenetic studies in Carnivora.
In Comparative Mammalian Cytogenetics (Benirschke, K., ed.), Springer-Verlag,
N. Y., 1969.

Lycaon pictus (African or Cape hunting dog)

2n=78

Order: CARNIVORA

Family: CANIDAE

Urocyon cinereoargenteus (Eastern gray fox)
$2n = 66$

Urocyon cinereoargenteus (Eastern gray fox)

2n=66

AUTOSOMES: 2 Metacentrics
             62 Acrocentrics

SEX CHROMOSOMES: X Submetacentric
                  Y Metacentric

A skin biopsy from a male was donated by Dr. H. Heck, Catskill Game Farm, Catskill, New York, USA, and that of the female, by Dr. C. Gray, National Zoological Park, Washington, D.C., USA. The X chromosomes were identified by radioautography. Two pairs of small acrocentrics have secondary constrictions.

REFERENCES:

1) Wurster, D.H. and Benirschke, K.: Comparative cytogenetic studies in the Order Carnivora. Chromosoma 24:336, 1968.

Urocyon cinereoargenteus (Eastern gray fox)

2n=66

X Y

X X

Order: CARNIVORA

Family: URSIDAE

Helarctos malayanus (Sun bear)
2n = 74

Order: CARNIVORA   Family: URSIDAE

<u>Helarctos malayanus</u> (Sun bear)

2n=74

AUTOSOMES: 14 Metacentrics and submetacentrics
             58 Acrocentrics

SEX CHROMOSOMES: X Submetacentric
                   Y Acrocentric

     Skin biopsies of one male and one female sun bear were kindly supplied
by Dr. C. Gray, the National Zoological Park, Washington, D.C., USA. They
gave similar results and the karyotype is similar to that of other Northern
hemisphere bears. The second largest acrocentric autosomal pair often shows
a secondary constriction near the centromere.

REFERENCES:

1) Wurster, D.H.: Cytogenetic and phylogenetic studies in carnivora. In
<u>Comparative Mammalian Cytogenetics</u> (Benirschke, K., ed.), Springer-Verlag,
N. Y., 1969.

<u>Helarctos</u> <u>malayanus</u> (Sun bear)

2n=74

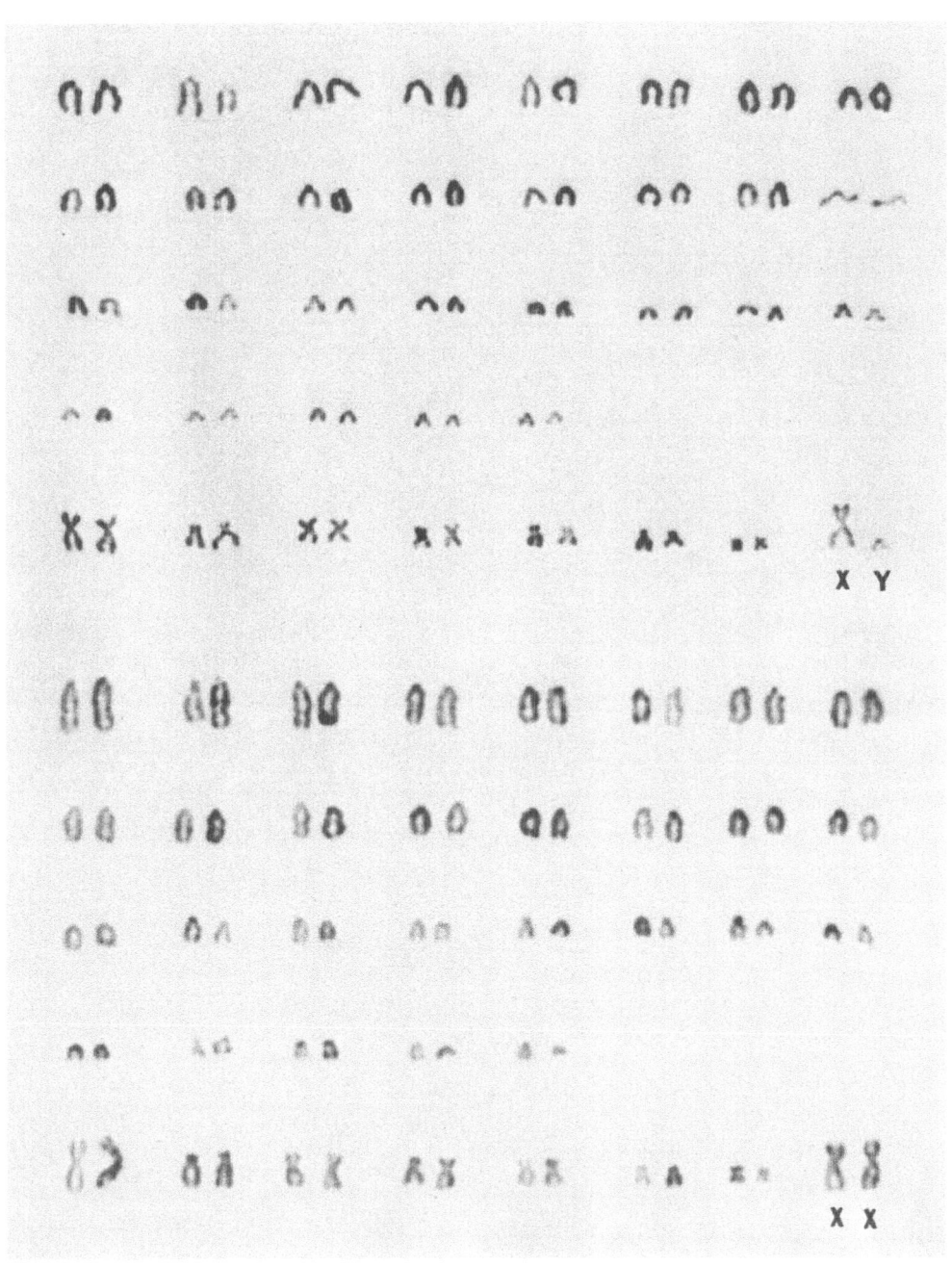

Order: CARNIVORA

Family: PROCYONIDAE

Ailurus fulgens (Lesser panda)

2n = 36

Order: CARNIVORA   Family: PROCYONIDAE

<u>Ailurus</u> <u>fulgens</u> (Lesser panda)

2n=36

AUTOSOMES: 32 Metacentrics and submetacentrics
                     2 Acrocentrics

SEX CHROMOSOMES: X Submetacentric
                                 Y Acrocentric

      Skin biopsies of one male and one female specimens were kindly supplied
by Dr. C. Gray, National Zoological Park, Washington, D.C., USA. They gave
similar results and agree with the findings of Todd and Pressman. Because
of the cytologic similarity to Felidae the Karyotype has been arranged like
that of cats. The prominent satellited marker chromosomes are placed in
the fourteenth position.

REFERENCES:

1) Wurster, D.H.: Cytogenetic and phylogenetic studies in carnivora.
In <u>Comparative Mammalian Cytogenetics</u> (Benirschke, K., ed.), Springer-Verlag,
N. Y., 1969.

2) Todd, N.B. and Pressman, S.R.: The karyotype of the lesser panda
(<u>Ailurus</u> <u>fulgens</u>) and general remarks on the phylogeny and affinities of
the panda. Carnivore Genetics Newsletter No. 5:105, 1968.

<u>Ailurus</u> <u>fulgens</u> (Lesser panda)

2n=36

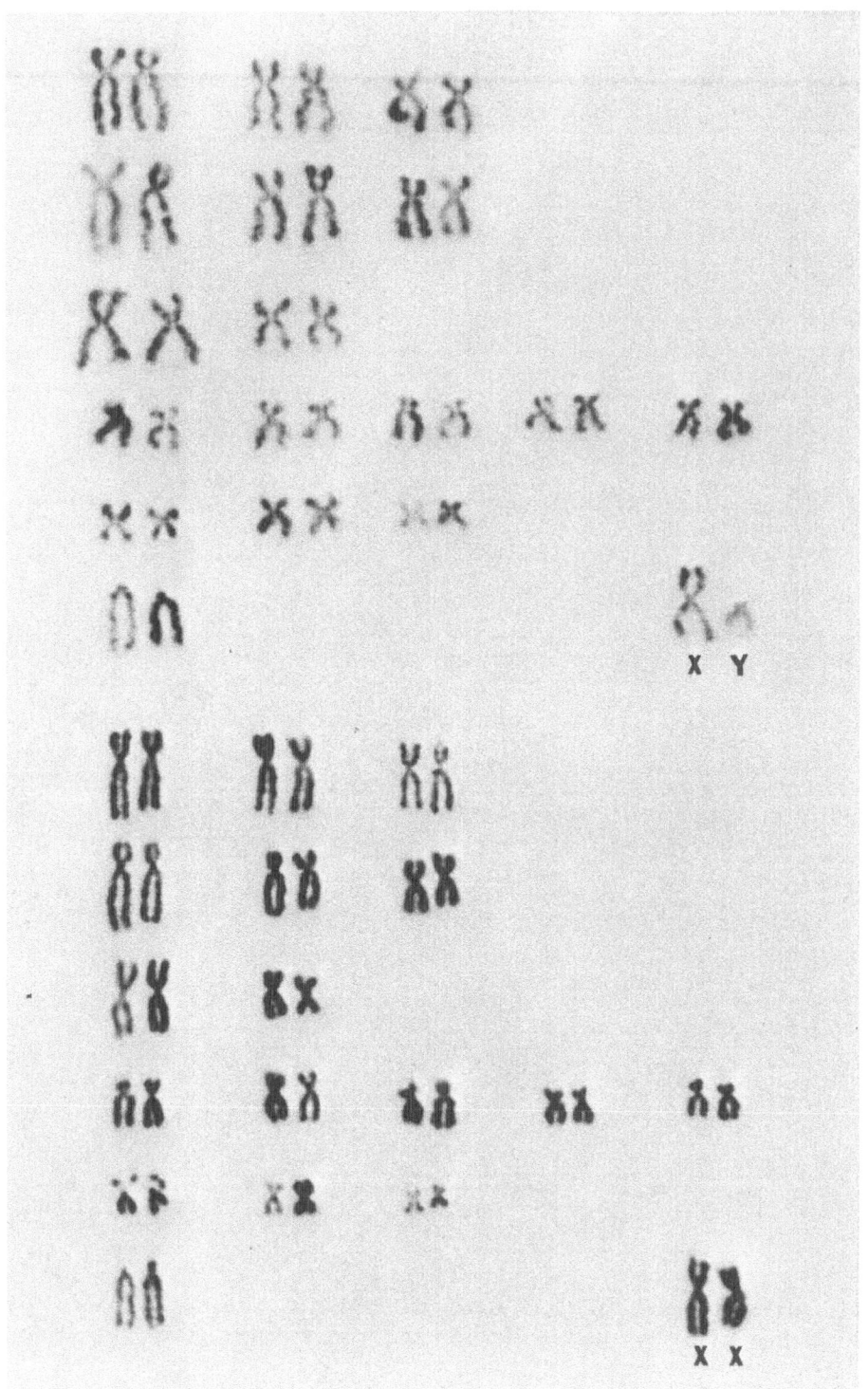

Order: CARNIVORA

Family: PROCYONIDAE

Nasua nasua (narica) (Coatimundi)
$2n = 38$

<u>Nasua</u> <u>nasua</u> (<u>narica</u>) (Coatimundi)

2n=38

AUTOSOMES: 30 Metacentrics and submetacentrics
             6 Acrocentrics

SEX CHROMOSOMES: X Submetacentric
                   Y Acrocentric

     Two male and two female animals were available for skin biopsy.
Karyotypes from cultured cells agree well with one another. A small
satellited pair of autosomes, placed first of the autosomes in row 4,
is the marker element.

REFERENCES:

1) Hsu, T.C. and Arrighi, F.E.: Karyotypes of 13 carnivores. Mammalian
Chromosomes Newsletter No. 21:155, 1966.

2) Todd, N.B., York, R.N. and Pressman, S.R.: The karyotypes of the
raccoon (<u>Procyon</u> <u>lotor</u> L.), Coatimundi (<u>Nasua</u> <u>narica</u> L.) and kinkajou
(<u>Potos</u> <u>flavus</u> Schreber). Mammalian Chromosomes Newsletter No. 21:153, 1966.

3) Panzetta, P. and Alaimo, I.: Karyotype of the coati (<u>Nasua</u> <u>nasua</u>
<u>solitaria</u> Schinz). Mammalian Chromosomes Newsletter <u>8</u>:97, 1967.

4) Wurster, D.H. and Benirschke, K.: Chromosome numbers in thirty species
of carnivores. Mammalian Chromosomes Newsletter <u>8</u>:195, 1967.

5) Wurster, D.H. and Benirschke, K.: Comparative cytogenetic studies in
the order <u>Carnivora</u>. Chromosoma <u>24</u>:336, 1968.

Nasua nasua (narica) (Coatimundi)

2n=38

Order: CARNIVORA

Family: MUSTELIDAE

Gulo gulo (Wolverine)
$2n = 42$

Gulo gulo (Wolverine)

2n=42

AUTOSOMES: 24 Metacentrics and submetacentrics
             16 Acrocentrics

SEX CHROMOSOMES: X Metacentric
                   Y Acrocentric

   Skin biopsies of two male and two female animals from the Yukon
Territory were made available through the kindness of the staff of the
University of Connecticut, Storrs, Connecticut, USA, and a dealer in
Vermont, USA. Other than the sex chromosomes, karyotypes of all specimens
agree with one another and agree with the findings of Fredga. The two
marker chromosomes with secondary constrictions in the long arms are placed
last of the autosomes.

REFERENCES:

1) Fredga, K.: Comparative chromosome studies in the family Mustelidae
(Carnivora-Mammalia). Hereditas 57:295, 1967.

2) Wurster, D.H. and Benirschke, K.: Chromosome numbers in thirty species
of carnivores. Mammalian Chromosomes Newsletter 8:195, 1967.

3) Wurster, D.H. and Benirschke, K.: Comparative cytogenetic studies in
the order Carnivora. Chromosoma 24:336, 1968.

Order: CARNIVORA

Family: MUSTELIDAE

Gulo gulo (Wolverine)

2n=42

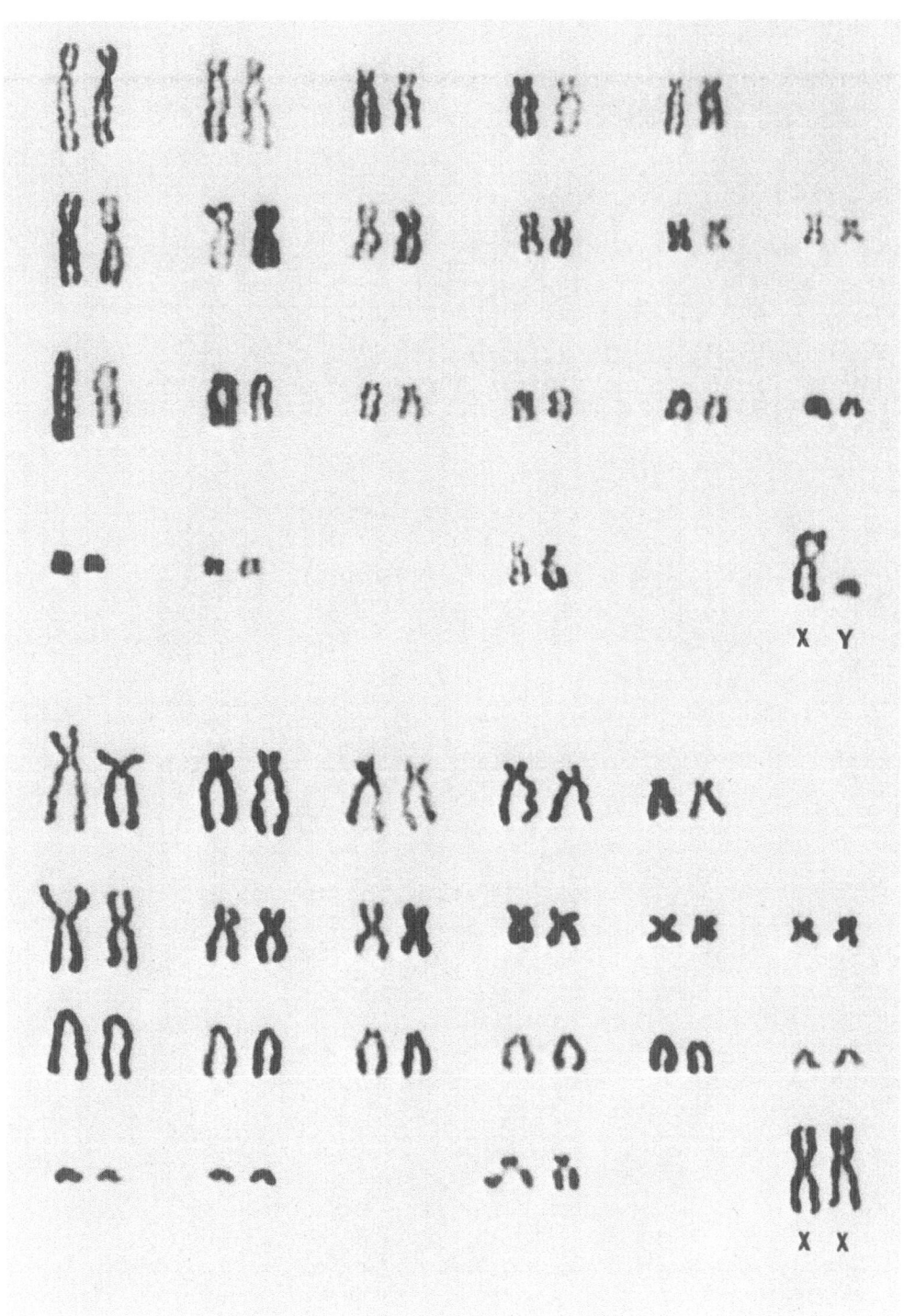

Order: CARNIVORA

Family: VIVERRIDAE

Herpestes auropunctatus (Indian mongoose)

$2n = 35 \, \male, \; 36 \, \female$

<u>Herpestes auropunctatus</u> (Indian mongoose)

2n=35♂,36♀

AUTOSOMES: 30 Metacentrics and submetacentrics
            4 Acrocentrics

SEX CHROMOSOMES: X Metacentric
                  Y Nonexistent as such

    The karyotypes were prepared from skin biopsies of one male and one female specimen and were kindly donated by Dr. M. M. Cohen, University of Buffalo, New York, USA. The X chromosomes were identified by radioautography, but the Y chromosome has not been identified. There is no marker chromosome. The taxonomic identification of animals studied by some authors is somewhat dubious (see ref. 6).

    Fredga (1967) found satellite associations among the acrocentric autosomes. He also presented an idiogram for this species and suggested that the Y was a minute and was translocated onto the satellite area of one of the acrocentric autosomes.

REFERENCES:

1) Fredga, K.: A new sex determining mechanism in a mammal. Chromosome of an Indian mongoose (<u>Herpestes auropunctatus</u>). Hereditas <u>52</u>:411, 1964.

2) Fredga, K.: New sex determining mechanism in mammal. Nature <u>206</u>: 1176, 1965.

3) Talukdar, M. and Manna, G.K.: Karyotypes of five carnivoran species from India. Mammalian Chromosomes Newsletter No. 21:151, 1966.

4) Manna, G.K. and Talukdar, M.: Somatic chromosome number in twenty species of mammals from India. Mammalian Chromosomes Newsletter No. 17:77, 1965.

5) Todd, N.B. and Pressman, S.R.: The karyotype of the lesser Indian mongoose (<u>Herpestes javanicus</u> Geoffroy), the meerkat (<u>Suricata suricatta</u> Desmarest) and comments on the taxonomy and karyology of the <u>Viverridae</u>. Mammalian Chromosomes Newsletter <u>8</u>:21, 1967.

6) Fredga, K.: Chromosome studies in six different tissues of a male Indian mongoose (<u>Herpestes auropunctatus</u>) and comments on the nomenclature of the species. Mammalian Chromosomes Newsletter <u>8</u>:19, 1967.

7) Cohen, M.M. and Chandra, H.S.: The somatic chromosomes of the small Indian mongoose: Autoradiographic analysis of an unbalanced translocation heterozygote. Cytogenetics (in press).

Herpestes auropunctatus (Indian mongoose)

2n=35♂, 36♀

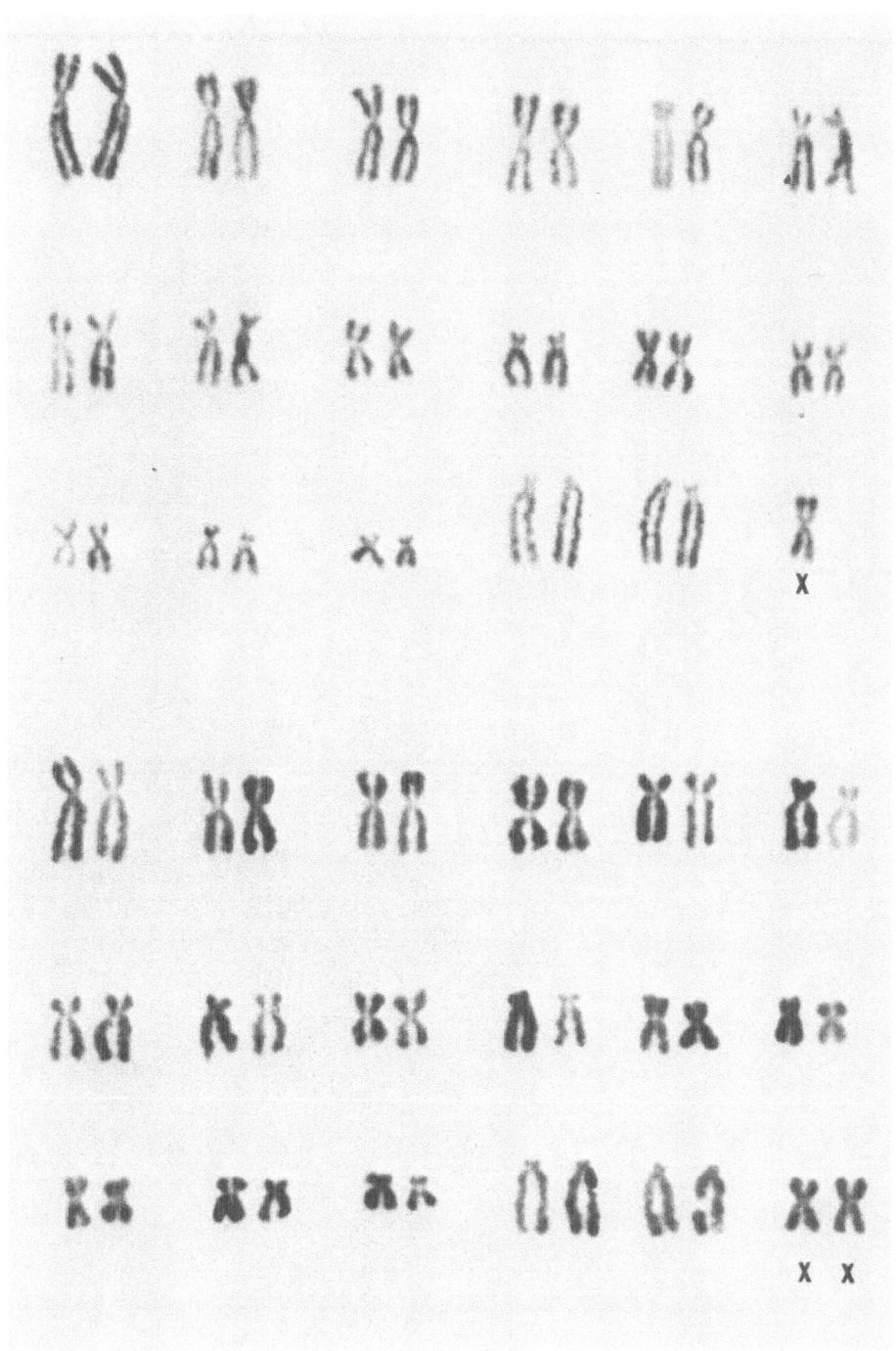

Order: CARNIVORA

Family: FELIDAE

Felis yagouaroundi (Jaguarundi)
2n = 38

Order: CARNIVORA   Family: FELIDAE

Felis yagouaroundi (Jaguarundi)

2n=38

AUTOSOMES: 36 Metacentrics

SEX CHROMOSOMES: X Submetacentric
                   Y Metacentric

    Skin biopsies of one male and one female animal were kindly provided
by Dr. C. Gray, National Zoological Park, Washington, D.C., USA. The
typical feline marker chromosomes are the first autosomes in the fourth row.

    It is of interest to note that in jaguarundi, all chromosomes are
biarmed yet the diploid number remains 38, the same as that of most cats,
which possess one or two pairs of acrocentrics.

REFERENCES:

1) Hsu, T. C., Arrighi, F.E. and Luquette, G.F.: Karyological studies of
nine species of Felidae. Amer. Naturalist 97:225, 1963.

2) Wurster, D.H. and Benirschke, K.: Comparative cytogenetic studies in
the Order Carnivora. Chromosoma 24:336, 1968.

3) Wurster, D.H.: Cytogenetic and phylogenetic studies in Carnivora. In
Comparative Mammalian Cytogenetics (Benirschke, K., ed.), Springer-Verlag,
N. Y., 1969.

4) Wurster, D. and Benirschke, K.: Karyotypes of four more species of cats.
Mammalian Chromosomes Newsletter 9:236, 1968.

Felis yagouaroundi (Jaguarundi)

2n=38

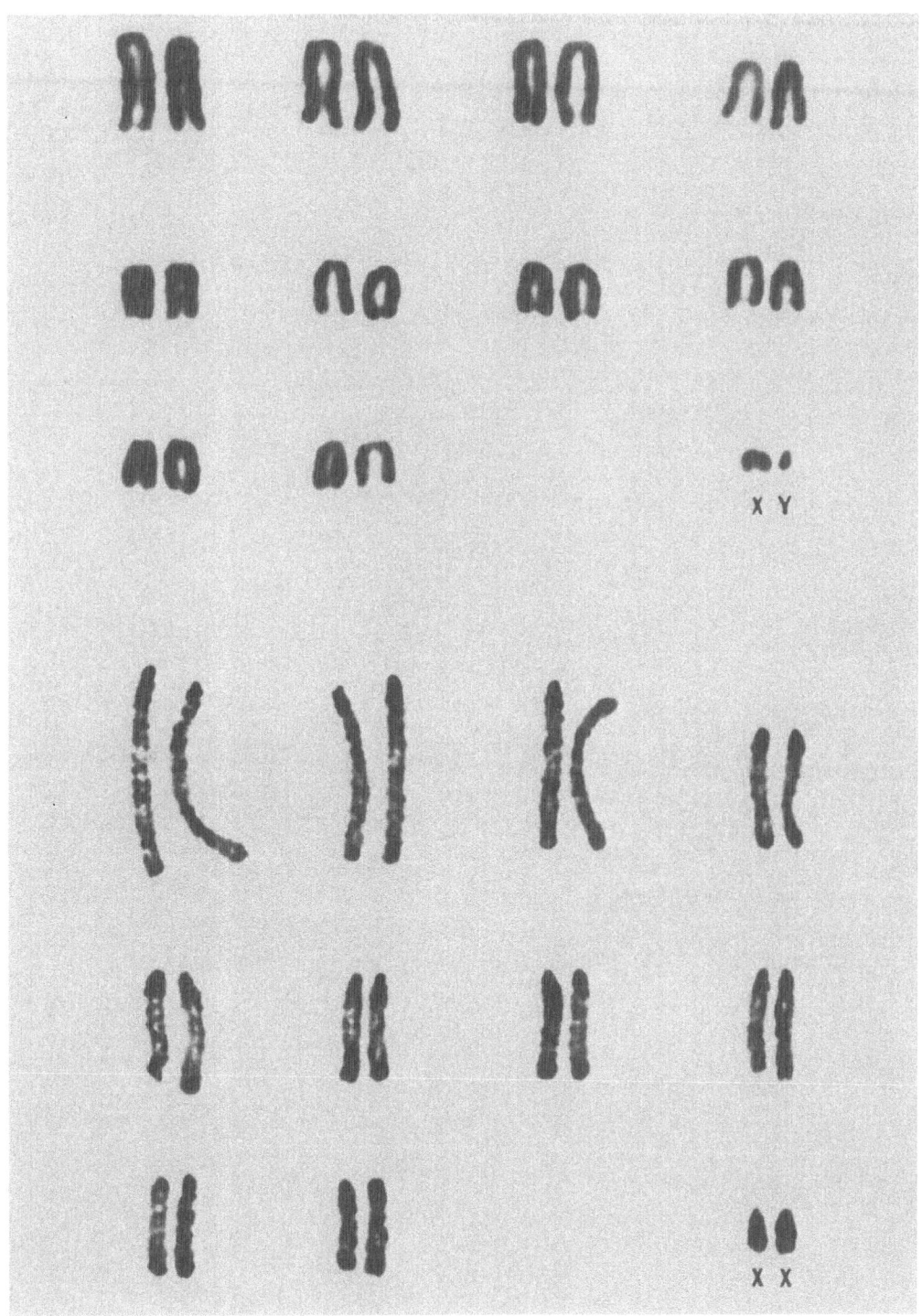

Order: CARNIVORA

Family: FELIDAE

Lynx rufus (Bobcat)

$2n = 38$

Order: CARNIVORA   Family: FELIDAE

<u>Lynx</u> <u>rufus</u> (Bobcat)

2n=38

AUTOSOMES: 32 Metacentrics, submetacentrics and subtelocentrics
            4 Acrocentrics

SEX CHROMOSOMES: X Submetacentric
                  Y Submetacentric

     The karyotype of the bobcat is very similar to that of the leopard
(Folio 84) except that its Y chromosome is slightly longer. It is also
very similar to one of its close relatives, the caracal, except that the
latter has a very large satellite on chromosome E1.

     The specimens were display animals from the Houston Zoological Garden,
Houston, Texas, USA. The origin of the animals is not available.

REFERENCES:

1) Hsu, T.C., Rearden, H.H. and Luquette, G.F.: Karyological studies of
nine species of Felidae. Amer. Nat. 97:225, 1963.

2) Sutton, D.A.: Karyotype of a female bobcat, <u>Lynx</u> <u>rufus</u> <u>californicus</u>.
Mammalian Chromosomes Newsletter <u>9</u>:249, 1968.

Order: CARNIVORA

Lynx rufus (Bobcat)

2n=38

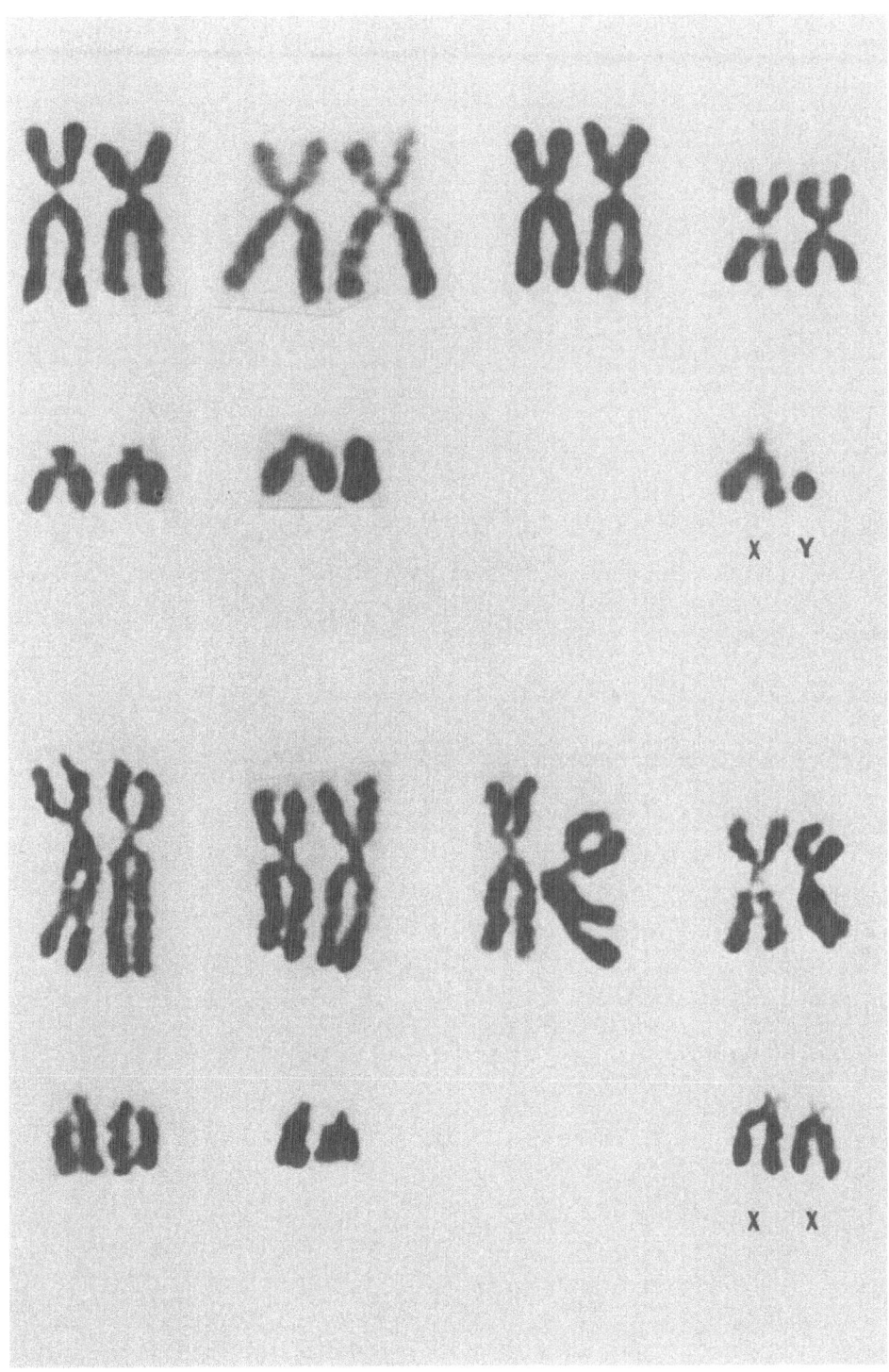

Order: ARTIODACTYLA

Family: BOVIDAE

Aepyceros melampus melampus (Impala)
2n = 60

Volume 4, Folio 188, 1970

<u>Aepyceros</u> <u>melampus</u> <u>melampus</u> (Impala)

2n=60

AUTOSOMES: 58 Acrocentrics

SEX CHROMOSOMES: X Acrocentric
                  Y Metacentric

Skin biopsies of two specimens, one male and one female, were made available by Dr. H. Heck, Catskill Game Farm, New York, USA. The karyotypes are essentially similar to those of Wallace and Fairall. These authors studied 34 animals (20♂, 14♀) of which 17 had 2n=60, 14 had 2n=59 (with one metacentric), and 3 had 2n=58 (with two metacentrics). These authors concluded that a Robertsonian system exists in this species and they found trivalent figures in meiosis of animals with 2n=59.

REFERENCES:

1) Wallace, C. and Fairall, N.: Chromosome polymorphism in the impala (<u>Aepyceros</u> <u>melampus</u> <u>melampus</u>). S. Afr. J. Sci. <u>63</u>:482, 1967.

2) Wurster, D.H. and Benirschke, K.: The chromosomes of twenty-three species of Cervoidea and Bovoidea. Mammalian Chromosomes Newsletter <u>8</u>: 226, 1967.

3) Wurster, D.H. and Benirschke, K.: Chromosome studies in the superfamily Bovoidea. Chromosoma <u>25</u>:152, 1968.

Aepyceros melampus melampus (Impala)

2n=60

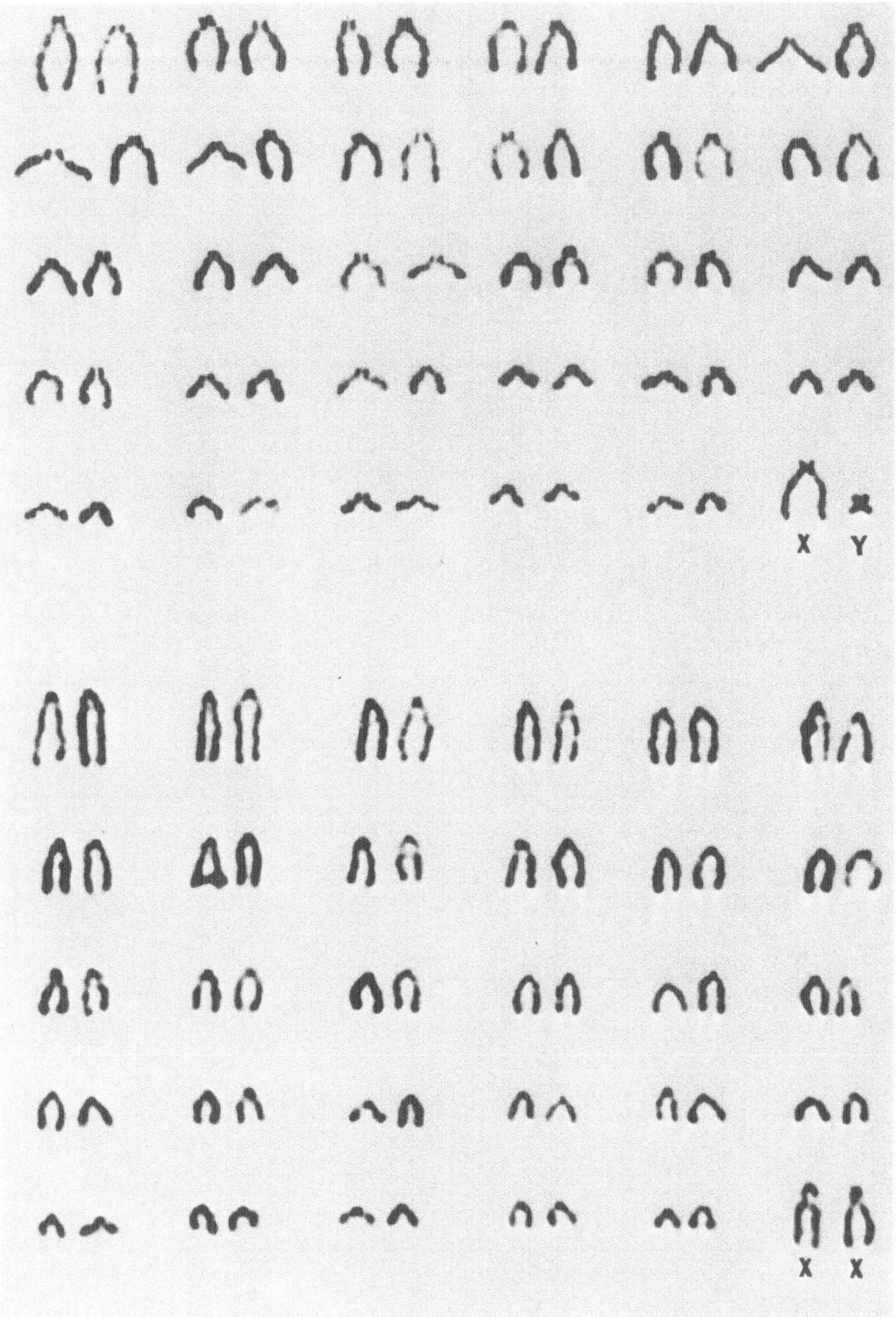

Order: ARTIODACTYLA

Family: BOVIDAE

Bubalus bubalis (Murrah buffalo)

$2n = 50$

Order: ARTIODACTYLA   Family: BOVIDAE

<u>Bubalus bubalis</u> (Murrah buffalo)

2n=50

AUTOSOMES: 10 Metacentrics and submetacentrics
             38 Acrocentrics

SEX CHROMOSOMES: X Acrocentric
                   Y Acrocentric

     These karyotypes were kindly donated by the late Dr. F. Ulbrich,
Giessen, Germany. They were made from lymphocyte cultures of animals in
Malaysia. The karyotype of the Murrah buffalo, an Indian dairy breed,
differs from that of the Asiatic swamp buffalo of Thailand (2n=48) in that
the former has two more acrocentrics which are presumably translocated onto
the short arms of No. 1 autosome of the latter. Hybrids are fertile and
have been studied cytogenetically (see Folio 139, Vol. 3).

REFERENCES:

1) Fischer, H. and Ulbrich, F.: Chromosomes of the Murrah buffalo and its
crossbreeds with the Asiatic swamp buffalo (<u>Bubalus bubalis</u>). Z. Tierz.
Züchtbiol. <u>84</u>:110, 1968.

Bubalus bubalis (Murrah buffalo)

2n=50

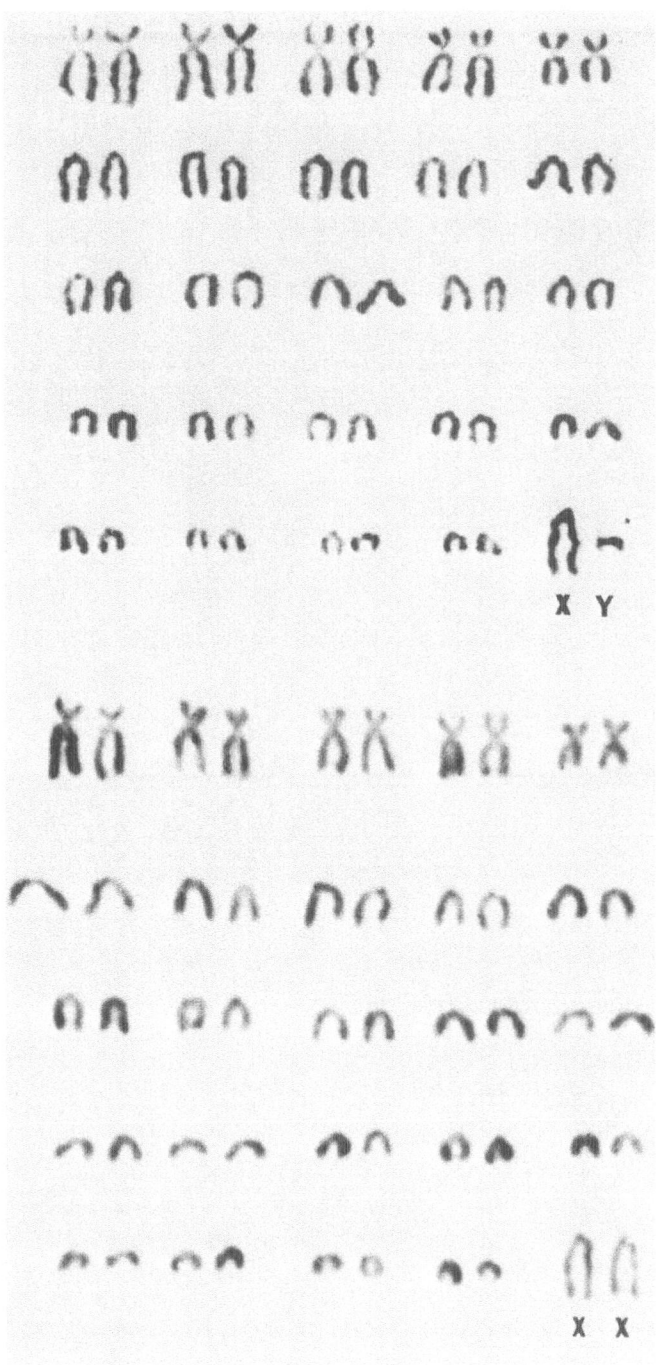

Order: ARTIODACTYLA

Family: BOVIDAE

Oreamnos americanus (Rocky Mountain goat)
$2n = 42$

Volume 4, Folio 190, 1970

Order: ARTIODACTYLA   Family: BOVIDAE

<u>Oreamnos</u> <u>americanus</u> (Rocky Mountain goat)

2n=42

AUTOSOMES: 18 Metacentrics and submetacentrics
             22 Acrocentrics

SEX CHROMOSOMES: X Acrocentric
                   Y Metacentric

     Skin biopsies from which these karyotypes were prepared came from one
male and two female animals shot in Montana, USA. They were kindly supplied
by Dr. B. W. O'Gara, Missoula, Montana. All specimens agree well in
karyotypes. The sex chromosomes are unequivocally identifiable.

REFERENCES:

1) Wurster, DH. and Benirschke, K.: The chromosomes of the Rocky Mountain
goat (<u>Oreamnos</u> <u>americanus</u>). Mammalian Chromosomes Newsletter <u>9</u>:80, 1968.

2) Wurster, D.H. and Benirschke, K.: Chromosome studies in the superfamily
Bovoidea. Chromosoma <u>25</u>:152, 1968.

Oreamnos americanus (Rocky Mountain goat)

2n=42

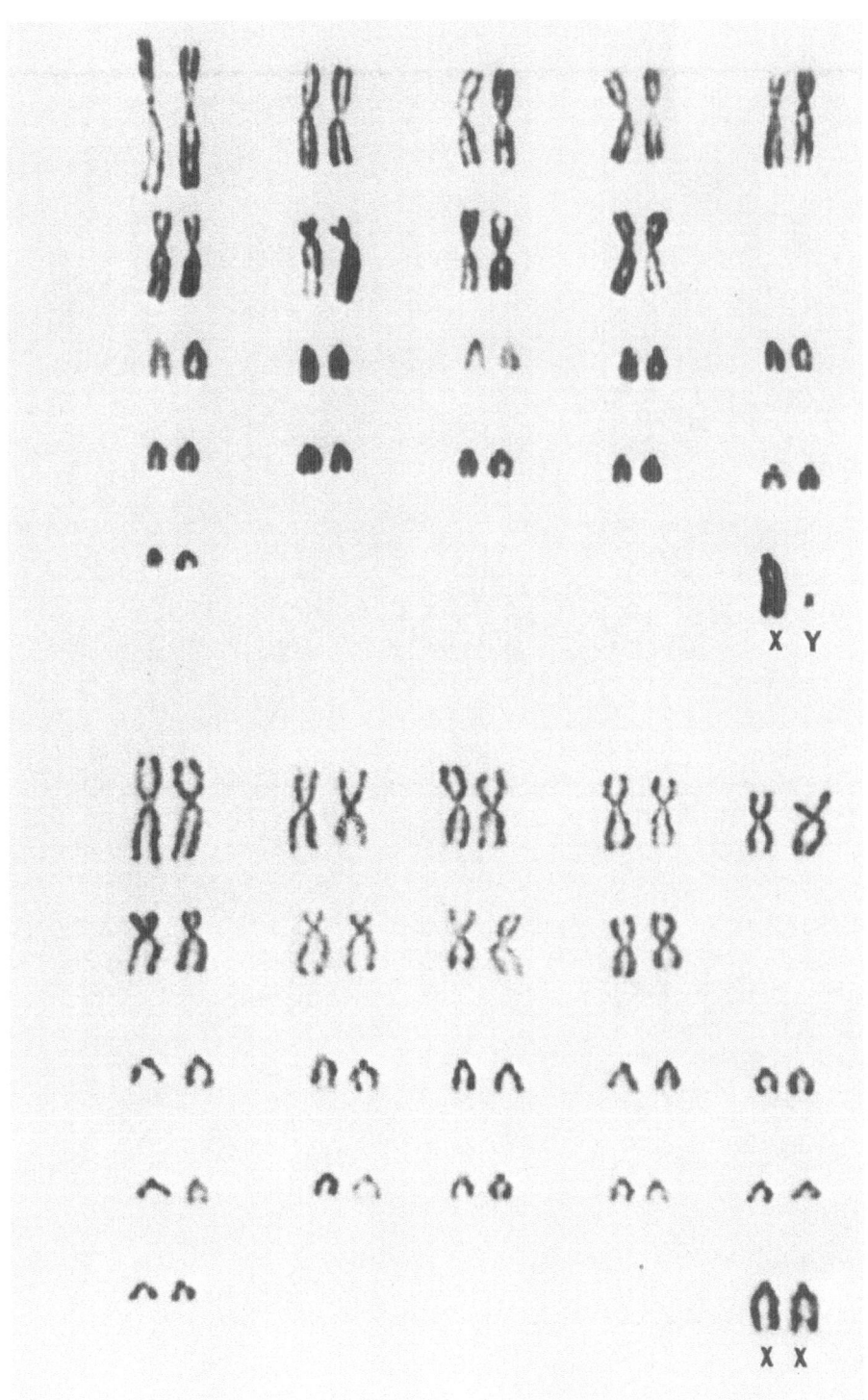

Order: ARTIODACTYLA

Family: BOVIDAE

Ovis ammon (aries) cycloceros (Afghanistan wild sheep)
2n = 58

Ovis ammon (aries) cycloceros (Afghanistan wild sheep)

2n=58

AUTOSOMES: 2 Metacentrics or submetacentrics
              54 Acrocentrics

SEX CHROMOSOMES: X Acrocentric
                  Y Metacentric

    These karyotypes came from blood cultures and were donated by the late
Dr. F. Ulbrich, Giessen, Germany. They are indistinguishable from those of
the aoudad, Ammotragus lervia (Folio 137, Vol. 3). A typical Robertsonian
system exists among members of the Caprinae with Capra hircus (Folio 92,
Vol. 2) and Capra ibex (Folio 140, Vol. 3) having only acrocentrics
(2n=60), the present species with 2 metacentrics (2n=58), Ovis a. nigrimontana
(Folio 144, Vol. 3) with 4 metacentrics (2n=56), Ovis aries (Folio 45, Vol. 1)
with 6 metacentrics (2n=54), and Hemitragus jemlahicus (Folio 141, Vol. 3)
with 12 metacentrics (2n=48).

REFERENCES:

1) Schmitt, J. and Ulbrich, F.: Die Chromosomen verschiedener Caprini
(Simpson, 1945). Z. Saugetierk. 33:180, 1968.

Ovis ammon (aries) cycloceros (Afghanistan wild sheep)

2n=58

Order: ARTIODACTYLA

Family: BOVIDAE

Syncerus caffer nanus (Congo buffalo)
2n = 54

Volume 4, Folio 192, 1970

Order: ARTIODACTYLA   Family: BOVIDAE

<u>Syncerus</u> <u>caffer</u> <u>nanus</u> (Congo buffalo)

2n=54

AUTOSOMES: 6 Metacentrics and submetacentrics
             46 Acrocentrics

SEX CHROMOSOMES: X Acrocentric
                  Y Acrocentric

     Skin biopsies of one male and one female animal were kindly made
available by Dr. H. Heck, Catskill Game Farm, New York, USA. This species
differs from the larger African buffalo, <u>Syncerus</u> <u>caffer</u> <u>caffer</u>,
(Folio 145, Vol. 3), in that it has two fewer metacentrics and four more
acrocentrics. A Robertsonian system is likely with two acrocentrics making
up the third pair of metacentrics in <u>Syncerus</u> <u>c</u>. <u>caffer</u>.

REFERENCES:

1) Wurster, D.H. and Benirschke, K.: The chromosomes of twenty-three
species of Cervoidea and Bovoidea. Mammalian Chromosomes Newsletter <u>8</u>:
226, 1967.

2) Heck, H., Wurster, D. and Benirschke, K.: Chromosome study of members
of the subfamilies Caprinae and Bovinae, family Bovidae: the Musk Ox, Ibex,
Aoudad, Congo Buffalo, and Gaur. Z. Saugetierk. <u>33</u>:172, 1968.

Syncerus caffer nanus (Congo buffalo)

2n=54

Order: PRIMATES

Family: TUPAIIDAE

Tupaia montana
$2n = 68$

Order: PRIMATES   Family: TUPAIIDAE

<u>Tupaia</u> <u>montana</u>

2n=68

AUTOSOMES: 4 Metacentrics and submetacentrics
             62 Acrocentrics

SEX CHROMOSOMES: X Large metacentric
                   Y Acrocentric

     Identification of the X chromosome is easy, but identification of the
Y chromosome is subjective. The two pairs of the biarmed autosomes are also
easy to recognize. One of the small acrocentric pairs showed secondary
constriction near its centromere.

     These karyotypes were from specimens collected on Mt. Kinabalu,
North Borneo in 1965 by Dr. C. H. Conaway, University of Missouri, Columbia,
Missouri, USA. Lung biopsies were used to initiate cultures for cytological
preparations.

REFERENCES:

1) Arrighi, F.E., Sorenson, M.W. and Shirley, L.R.: Chromosomes of the tree
shrews (<u>Tupaiidae</u>). Cytogenetics <u>8</u>:199, 1969.

Tupaia montana

2n=68

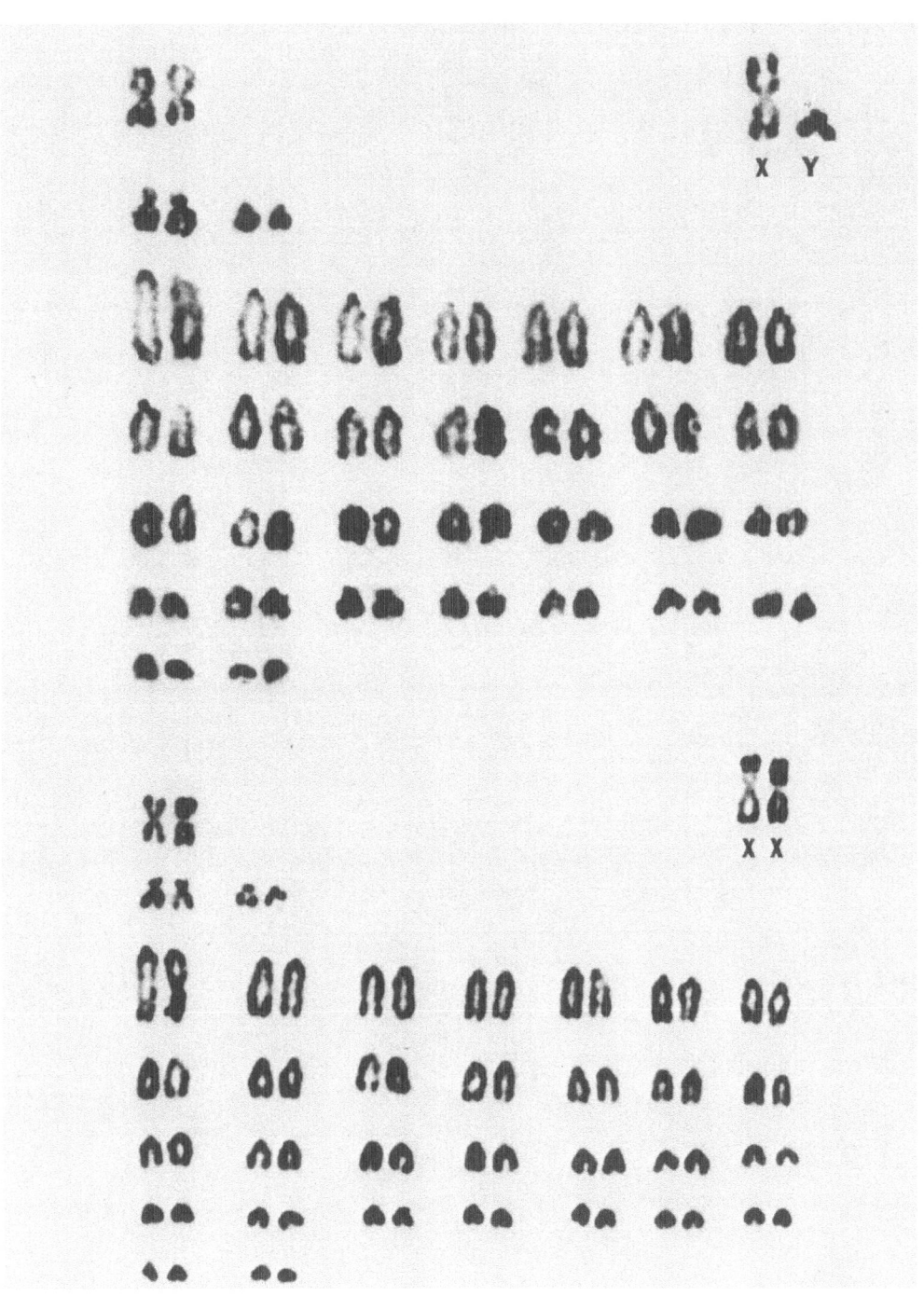

Order: PRIMATES

Family: LEMURIDAE

Lemur catta (Ring-tailed lemur)
$2n = 56$

Lemur catta (Ring-tailed lemur)

2n=56

AUTOSOMES: 8 Metacentrics and submetacentrics

SEX CHROMOSOMES: X Submetacentric
                   Y Acrocentric

One pair of the submetacentric autosomes is especially large. Several pairs of the acrocentric autosomes are very small elements. Two pairs of small acrocentric autosomes bear secondary constriction near the centromere.

The X chromosome is difficult to distinguish from the smallest pair submetacentric autosomes, and the determination of the Y is highly subjective.

The karyotypes are gifts of Dr. Ernest H.Y. Chu, Oak Ridge National Laboratory, Oak Ridge, Tennessee, USA. The specimens were originally collected by Dr. John Buettner-Janusch in Madagascar.

REFERENCES:

1) Chu, E.H.Y. and Swomley, B.A.: Chromosomes of lemurine lemurs. Science 133:1925, 1961.

2) Chiarelli, B.: Some chromosome numbers in Primates. Mammalian Chromosomes Newsletter No. 6,3, 1961.

3) Bender, M.A. and Chu, E.H.Y.: The chromosomes of Primates. In Evolutionary and Genetic Biology of Primates, Vol. 1 (J.Buettner-Janusch, ed.), Academic Press, N.Y., 1963.

4) Egozcue, J.: Chromosome variability in the Lemuridae. Amer. J. Phys. Anthropol. 26:341, 1967.

Order: PRIMATES

Family: LEMURIDAE

<u>Lemur</u> <u>catta</u> (Ring-tailed lemur)

2n=56

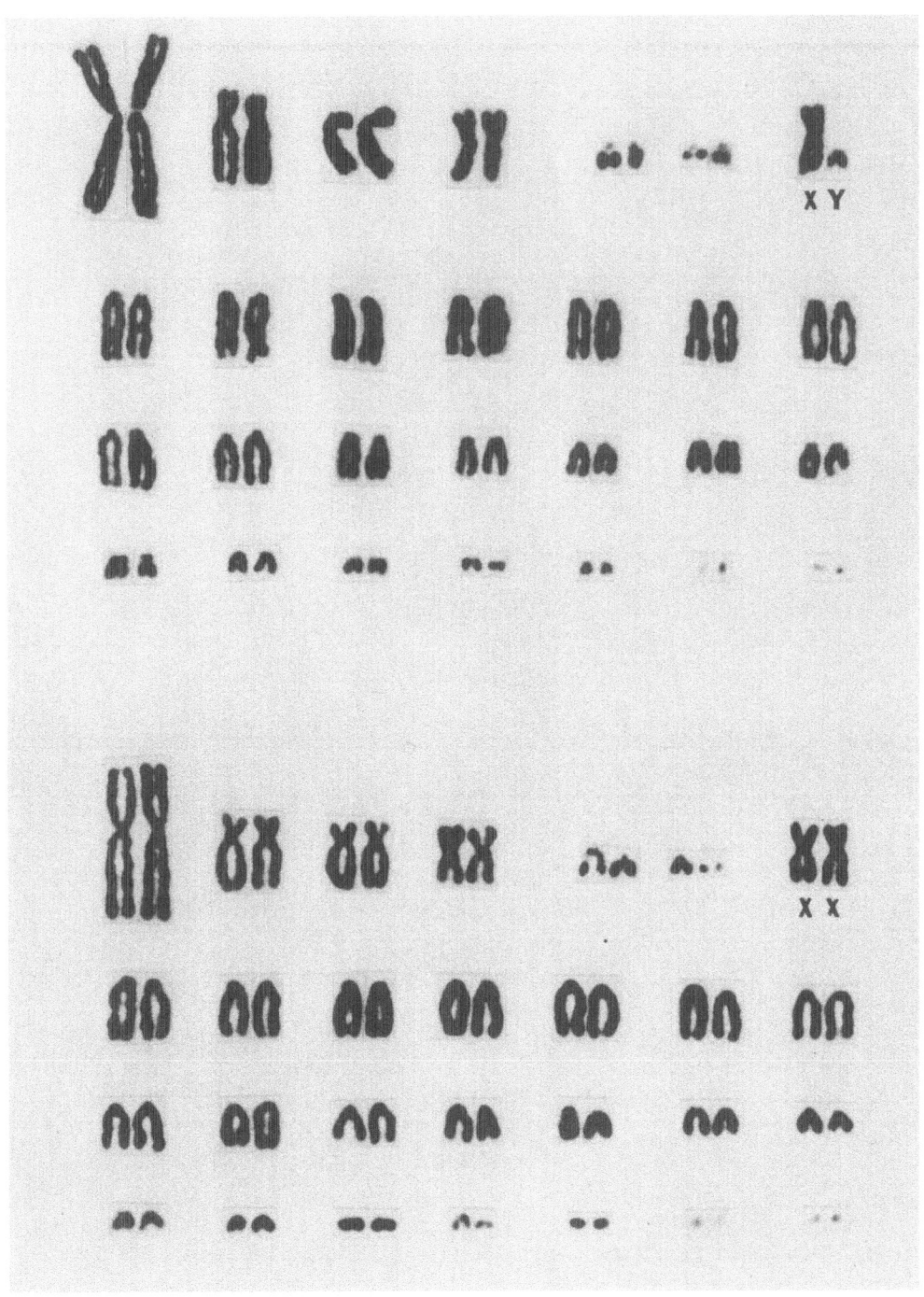

Order: PRIMATES

Family: CALLITHRICIDAE

Callithrix argentata (Silver marmoset)

$2n = 44$

<u>Callithrix</u> <u>argentata</u> (Silver marmoset)

2n=44

AUTOSOMES: 28 Metacentrics, submetacentrics and subtelocentrics
             14 Acrocentrics

SEX CHROMOSOMES: X Submetacentric
                   Y Minute acrocentric

    Several acrocentrics actually bear distinct second arms, so that they
may be classified as subtelocentrics. Secondary constrictions are not
obvious on any of the acrocentrics. Identification of the X is subjective
even in the male cells, but identification of the Y offers no problem.

    The karyotypes are from animals kept in captivity in The University of
Texas Dental Research Institute, Houston, Texas. A piece of ear fragment
was used from each specimen to initiate cell cultures for cytological
preparations.

REFERENCES:

1) Egozcue, J., Perkins, E.M. and Hagemenas, F.: Chromosomal evolution in
marmosets, tamarins and pinchés. Folia Primatol. <u>9</u>:81, 1968.

2) Egozcue, J.: Primates. In <u>Comparative</u> <u>Mammalian</u> <u>Cytogenetics</u>
(Benirschke, K., ed.), Springer-Verlag, New York, 1969.

Order: PRIMATES   Family: CALLITHRICIDAE

Callithrix argentata (Silver marmoset)

2n=44

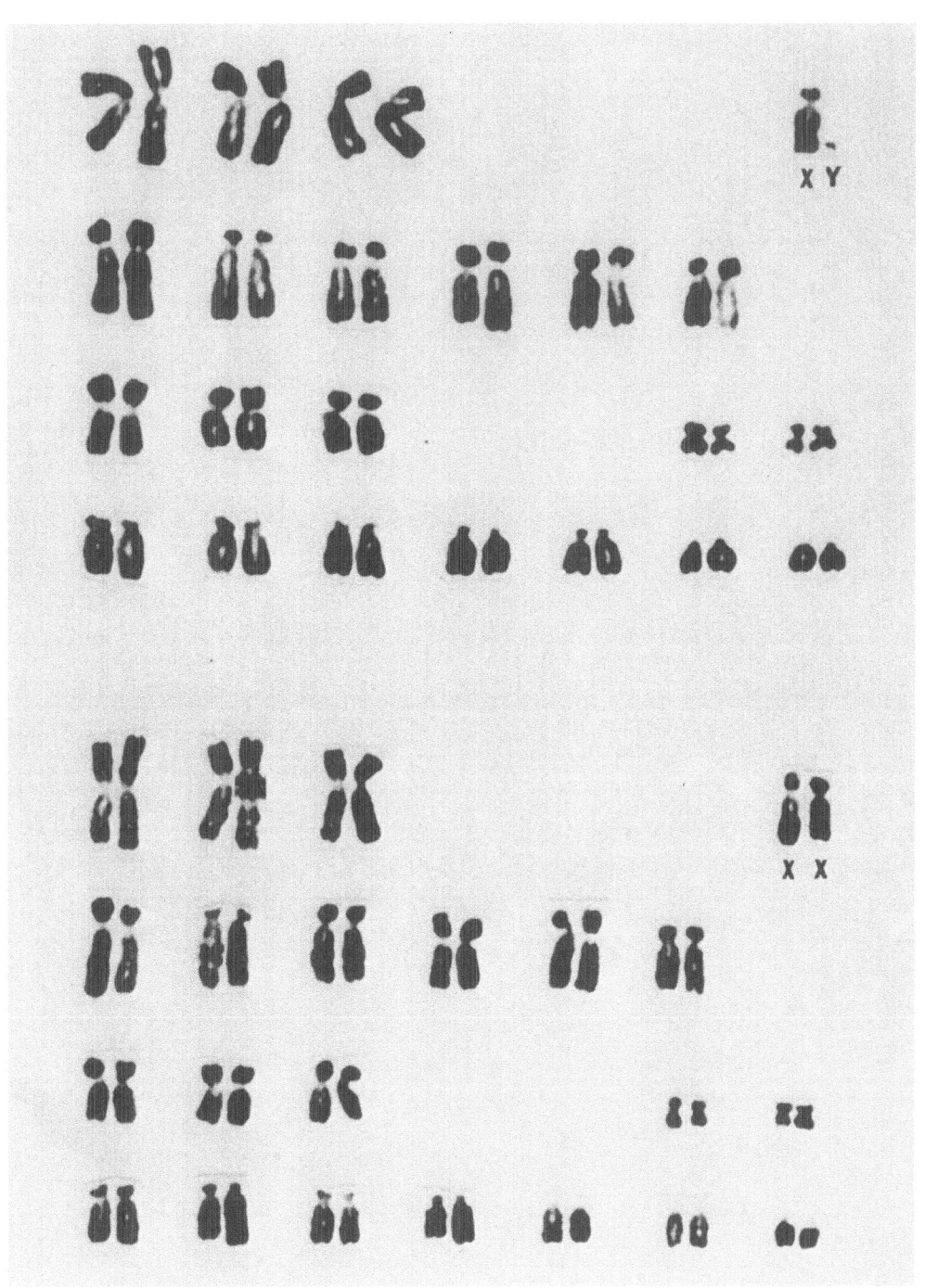

Order: PRIMATES

Family: CALLITHRICIDAE

Callithrix (Hapale) humeralifer (White-shouldered marmoset)

2n = 44

<u>Callithrix</u> (<u>Hapale</u>) <u>humeralifer</u> (White-shouldered marmoset)

2n=44

AUTOSOMES: 32 Metacentrics and submetacentrics
            10 Acrocentrics

SEX CHROMOSOMES: X Submetacentric
                  Y Acrocentric

These karyotypes came from lymphocyte cultures of one male and one female specimens and were kindly supplied by Dr. J. Egozcue, Oregon Regional Primate Center, Beaverton, Oregon, USA. The karyotype is similar to that of <u>Cebuella</u> <u>pygmaea</u> (Folio 47, Vol. 1) and <u>Saimiri</u>.

REFERENCES:

1) Egozcue, J., Perkins, E.M. and Hagemenas, F.: Chromosomal evolution in marmosets, tamarins and pinchés. Folia primatol. <u>9</u>:81, 1968.

2) Egozcue, J.: Primates. In <u>Comparative Mammalian Cytogenetics</u> (Benirschke, K., ed.), Springer-Verlag, N. Y., 1969.

Callithrix (Hapale) humeralifer (White-shouldered marmoset)

2n=44

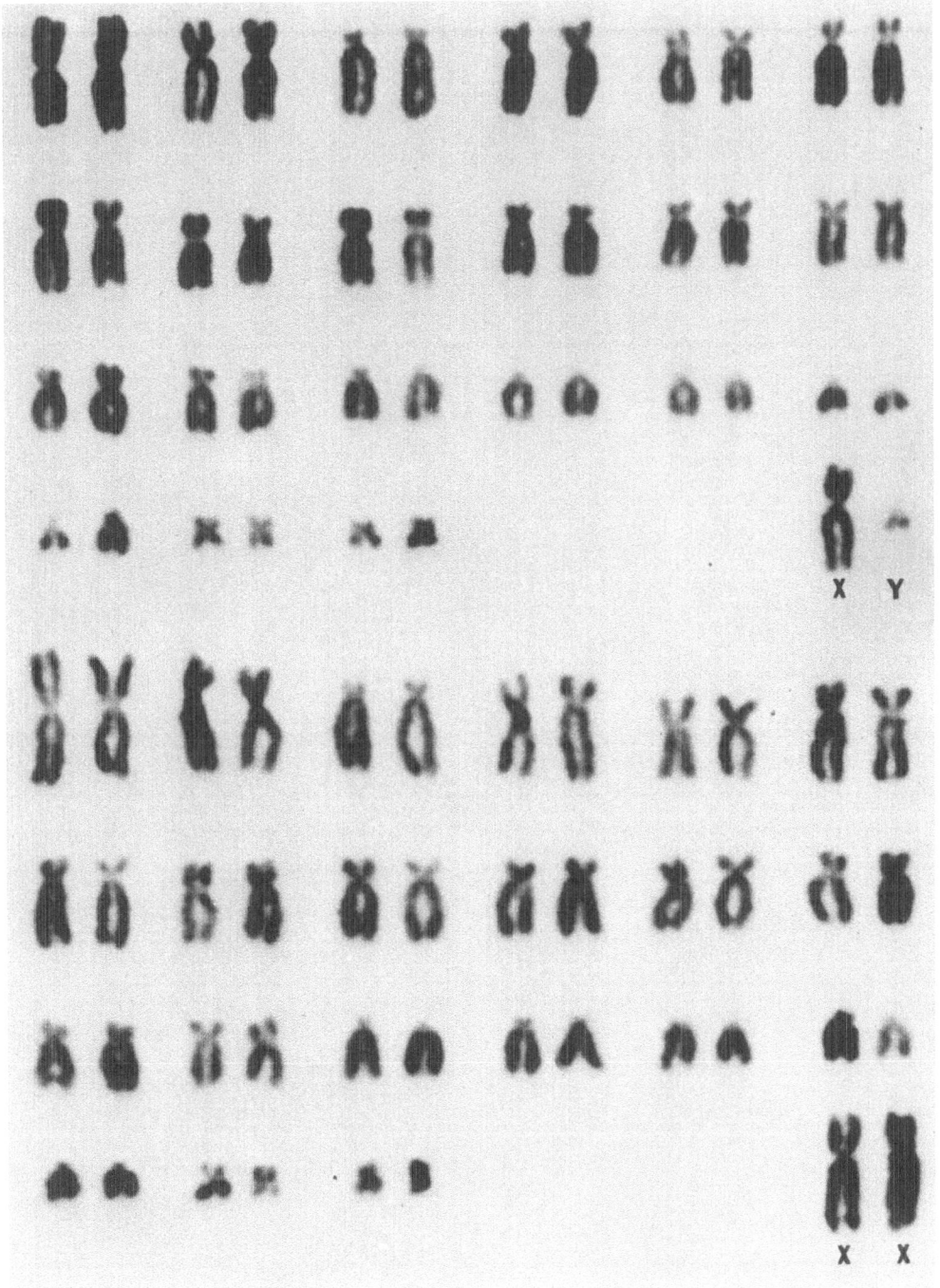

Order: PRIMATES

Family: CALLITHRICIDAE

Saguinus fuscicollis illigeri (Brown-headed tamarin)
$2n = 46$

Saguinus fuscicollis illigeri (Brown-headed tamarin)

2n=46

AUTOSOMES: 30 Metacentrics and submetacentrics
            14 Acrocentrics

SEX CHROMOSOMES: X Submetacentric
                  Y Metacentric

Two specimens, one male and one female, kept in the University of Texas Dental Research Institute at Houston were used for karyological studies. Fragments of ears were used to initiate cell cultures. The karyotypes presented here are similar to those of S. f. illigeri analyzed by Egozcue, who studied three males and three females. Karyologically this is also similar to the red-mantled tamarin by Bender and Chu and S. nigricollis (Folio 100, Vol. 2). Chimerism was found in blood and spermatogonia by Egozcue et al. Four acrocentrics have secondary constrictions, some have satellites.

The taxonomy of this group is still unsettled, and Egozcue followed the species designation of Hershkovitz (1966).

REFERENCES:

1) Bender, M.A. and Mettler, L.E.: Chromosome studies of Primates. II. Callithrix, Leontocebus and Callimico. Cytologia 25:400, 1960.

2) Egozcue, J., Perkins, E.M. and Hagemenas, F.: The chromosomes of Saguinus fuscicollis illigeri (Pucheran, 1845) and Aotus trivirgatus (Humboldt, 1811). Folia primatol. 10:154, 1969.

3) Hershkovitz, P.: Taxonomic notes on tamarins, genus Saguinus (Callithricidae, Primates), with descriptions of four new forms. Folia primatol. 4:381, 1966.

4) Egozcue, J.: Primates. In Comparative Mammalian Cytogenetics (Benirschke, K., ed.), Springer-Verlag, N. Y., 1969.

Order: PRIMATES   Family: CALLITHRICIDAE

Saguinus fuscicollis illigeri (Brown-headed tamarin)

2n=46

Order: PRIMATES

Family: CALLITHRICIDAE

Saguinus oedipus (Cottontop pinche)
$2n = 46$

<u>Saguinus oedipus</u> (Cottontop pinché)

2n=46

AUTOSOMES: 30 Metacentrics and submetacentrics
            14 Acrocentrics

SEX CHROMOSOMES: X Metacentric
                  Y Acrocentric

Two pairs of the acrocentric autosomes bear secondary constrictions near the centromeres. The sex chromosomes are easy to identify.

The animals were kept in captivity in the Dental Research Institute of the University of Texas Dental Branch at Houston, Houston, Texas, USA. A piece of ear fragment was used from each specimen to initiate cell cultures for cytological preparations.

REFERENCES:

1) Chiarelli, B.: Some chromosome numbers in primates. Mammalian Chromosomes Newsletter No. 6: 3, 1961.

2) Benirschke, K. and Brownhill, L.E.: Further observations on marrow chimerism in marmosets. Science 138:513, 1962.

3) Egozcue, J., Perkins, E.M. and Hagemenas, F.: Chromosomal evolution in marmosets, tamarins and pinchés. Folia primat. 9:81, 1968.

Order: PRIMATES

<u>Saguinus</u> <u>oedipus</u> (Cottontop pinché)

2n=46

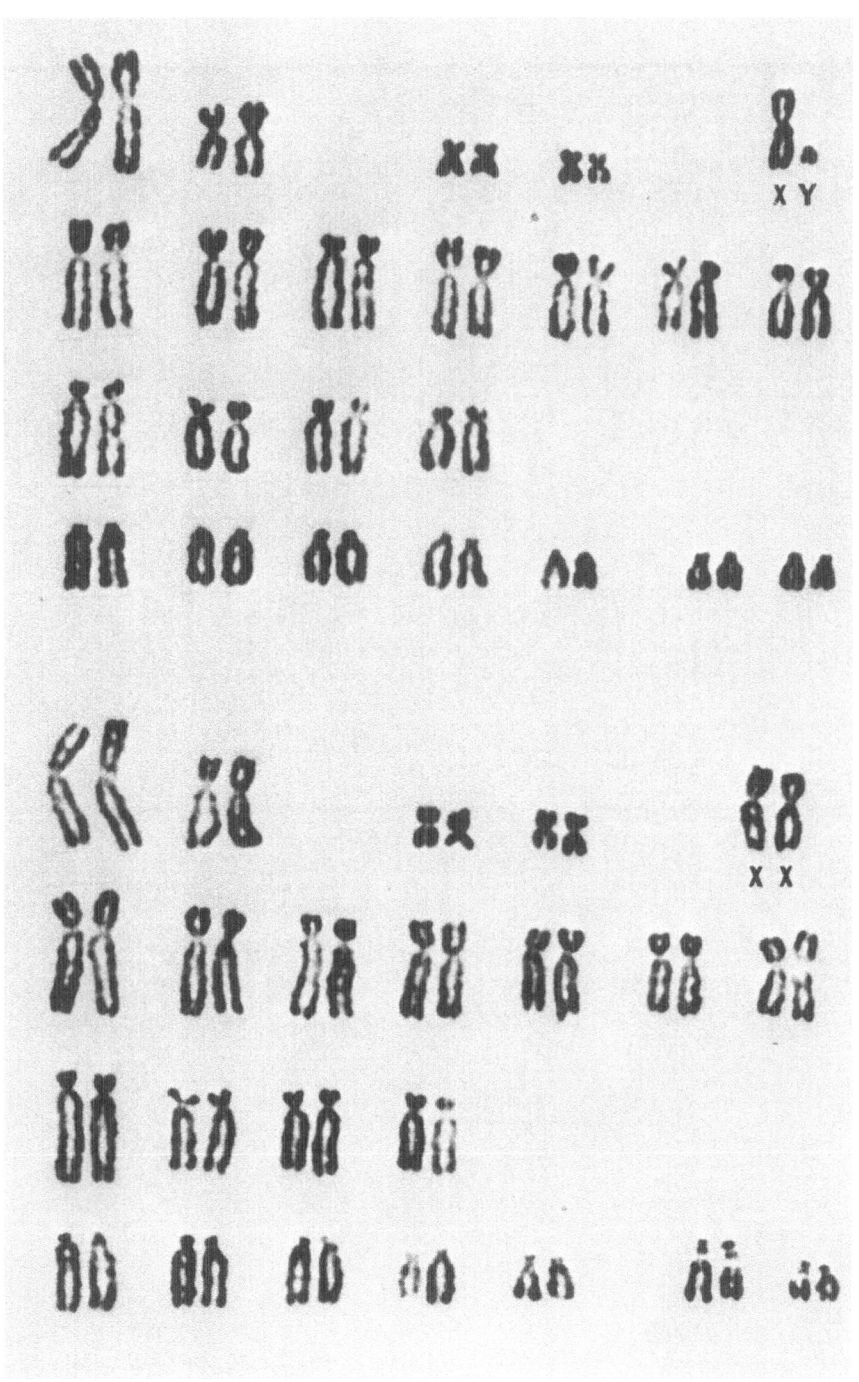

Order: PRIMATES

Family: COLOBIDAE

Presbytis cristatus (Silvered leaf-monkey)

$2n = 44$

Presbytis cristatus (Silvered leaf-monkey)

2n=44

AUTOSOMES: 40 Metacentrics and submetacentrics
             2 Acrocentrics

SEX CHROMOSOMES: X Submetacentric
                   Y Submetacentric

     Skin biopsies of two animals were made available by Dr. C. Gray,
National Zoological Park, Washington, D.C., USA. In contrast to other
langurs with 2n=44 (see Egozcue), all metaphases of this male have a larger
unmatched chromosome that appears to be the Y. This element is not
specifically identified as the Y chromosome, however. Pronounced secondary
constrictions are found in the long arms of one small metacentric chromosome,
here placed as the last autosome.

REFERENCES:

1) Wurster, D.H. and Benirschke, K.: Chromosomes of some primates.
Mammalian Chromosomes Newsletter 10:3, 1969.

2) Egozcue, J.: Primates. In Comparative Mammalian Cytogenetics
(Benirschke, K., ed.), Springer-Verlag, N. Y., 1969.

Presbytis cristatus (Silvered leaf-monkey)

2n=44

Order: PRIMATES

Family: PONGIDAE

Symphalangus brachytanites (Hylobates klossii) (Dwarf siamang)
2n = 50

Symphalangus brachytanites (Hylobates klossii) (Dwarf siamang)

2n=50

AUTOSOMES: 46 Metacentrics and submetacentrics
            2 Acrocentrics

SEX CHROMOSOMES: X Submetacentric
                  Y Metacentric

    Skin biopsies of one male and one female specimen, designated as
S. brachytanites, were kindly made available by Dr. J. Moor-Jankowski,
New York University Medical School, New York, USA. These animals are
"Dwarf Siamangs", Hylobates klossii according to Groves (1968), which
differ morphologically from the largest gibbon, S. syndactylus.
Karyotypically these animals are identical to the siamangs described
in the literature, except perhaps for the Y chromosome which is
submetacentric in S. syndactylus (Egozcue, 1969).

REFERENCES:

1) Bender, M.A. and Chu, E.H.Y.: The chromosomes of primates. In
Evolutionary and Genetic Biology of Primates (Buettner-Janusch, J., ed.),
Academic Press, N. Y., 1963.

2) Klinger, H.P.: The somatic chromosomes of some primates (Tupaia glis,
Nycticebus coucang, Tarsius bancanus, Cercocebus aterrimus, Symphalangus
syndactylus). Cytogenetics 2:140, 1963.

3) Wurster, D.H. and Benirschke, K.: Chromosomes of some primates.
Mammalian Chromosomes Newsletter 10:3, 1969.

4) Groves, C.P.: The classification of the gibbons (Primates, Pongidae).
Z. Säugetierk. 33:239, 1968.

5) Egozcue, J.: Primates. In Comparative Mammalian Cytogenetics
(Benirschke, K., ed.), Springer-Verlag, N. Y., 1969.

Symphalangus brachytanites (Hylobates klossii) (Dwarf siamang)

2n=50

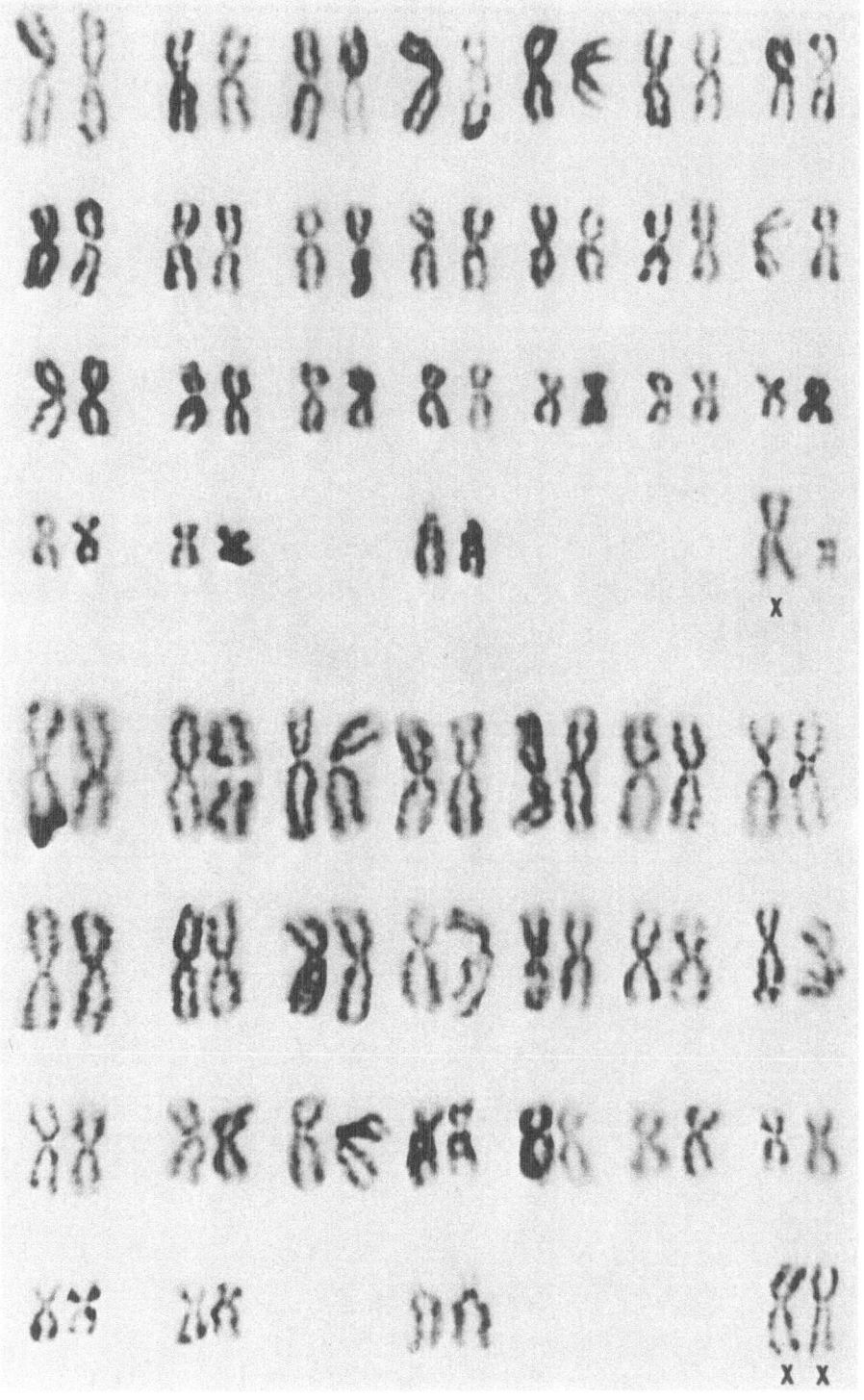

Cumulative Index   (Volumes 1 to 4)

Vernacular Names

Cumulative Index   (Volumes 1 to 4)

Technical Names

AN ATLAS OF
MAMMALIAN
CHROMOSOMES

VOLUME 5

AN ATLAS OF
MAMMALIAN
CHROMOSOMES

VOLUME 5

T. C. HSU

Section of Cytology, Department
of Biology, The University of
Texas M. D. Anderson Hospital and
Tumor Institute, Houston, Texas

KURT BENIRSCHKE

Department of Obstetrics
& Gynecology, School of Medicine,
University of California, San Diego,
La Jolla, California

SPRINGER SCIENCE+BUSINESS MEDIA, LLC 1971

© 1971 by Springer Science+Business Media New York
Originally published by Springer-Verlag New York Heidelberg Berlin in 1971
Softcover reprint of the hardcover 1st edition 1971

ISBN 978-1-4684-7386-5   ISBN 978-1-4615-6428-7 (eBook)
DOI 10.1007/978-1-4615-6428-7

Introduction

Since the inception of the Atlas of Mammalian Chromosomes, it is a surprise to us that we have finished five volumes with karyotypes of nearly 250 mammalian species. We acknowledge the fact that there are many imperfections in this series, but we also hope that these volumes have been useful to various investigators.

Springer-Verlag has made a binder which will accommodate the material of all five volumes. We also revised our index to facilitate easy reference. If the material is arranged in the binder according to the Cumulative Contents, we believe one will find it convenient to consult with this book.

Volume 5, as in the previous volumes, contains a Cumulative Index, Cumulative Table of Contents, as well as an individual Index and Table of Contents for Volume 5. Previous Indices and Tables of Contents should be discarded.

October, 1970

T. C. Hsu
Kurt Benirschke

Instructions

A special vinyl binder is available from the publisher which will accommodate the material of the first five volumes. Twenty dividers have been provided with printed tabs. These dividers are intended to separate the Orders and the Index. It is suggested that the dividers be placed in the following sequence of Orders, which sequence will be followed in the future when presenting karyotypes in new Orders:

Monotremata
Marsupialia
Insectivora
Dermoptera
Chiroptera
Edentata
Tubulidentata
Lagomorpha
Rodentia
Sirenia
Pholidota
Hyracoidea
Proboscidea
Carnivora
Pinnipedia
Cetacea
Perissodactyla
Artiodactyla
Primates
Index

These additional features should make it more convenient to use the Atlas in the future.

1. Folios should be arranged according to the Cumulative Table of Contents of Volumes 1-5, which is included in Volume 5. *If the Vinyl Binder is used, contents may be separated according to the taxonomic orders, each of which has an index plate for convenience.*

2. Use the Cumulative Index (Volumes 1-5) to find whether a particular species is included. If so, find the order and family of that particular species (abbreviations in parentheses) and follow the folio numbers.

3. The new Bibliographical References should, as in the past, be incorporated into the previous folios. These new references are printed on only one side of their respective sheets and are arranged in accordance with the order of this Atlas and folio numbers. They should be cut apart and pasted into the respective folio as indicated.

Contents, Volume 5

Cumulative Contents (Volumes 1 to 5)

New References for Previous Volumes

In Volume 5 there are 50 new species of various orders. These should be integrated with the Folios of Volumes 1, 2, 3 and 4. A new cumulative index is supplied, hence the old index should be discarded.

In addition to the new Folios, 5 pages of references for the Folios of Volumes 1, 2, 3 and 4 have been prepared to keep the Atlas up-to-date and useful. These references are arranged in such a fashion that they can be cut out and pasted into the previous Folios in proper sequence.

MARSUPIALIA

Vol. 1, Folio 1

12) Brinkley, B.R. and Humphrey, R.M.: Evidence for subchromatid organization in marsupial chromosomes. J. Cell Biol. $\underline{42}$:827, 1969.

INSECTIVORA

Vol. 2, Folio 52

10) Gropp, A., Citoler, P. and Geisler, M.: Karyotypvariation und Heterochromatinmuster bei Igeln (Erinaceus und Hemiechinus). Chromosoma $\underline{27}$: 288, 1969.

Vol. 2, Folio 53

9) Gropp, A., Citoler, P. and Geisler, M.: Karyotypvariation und Heterochromatinmuster bei Igeln (Erinaceus und Hemiechinus). Chromosoma $\underline{27}$: 288, 1969.

EDENTATA

Vol. 4, Folio 160

3) Corin-Frederic, J.: Les formules gonosomiques dites aberrantes chez les Mammifères Euthériens. Exemple particulier du paresseux Choloepus hoffmanni Peters (Edente, Xenarthre, famille des Bradypodidae). Chromosoma $\underline{27}$:268, 1969.

LAGOMORPHA

Vol. 1, Folio 8

18) Shaver, E.L. and Carr, D.H.: The chromosome complement of rabbit blastocysts in relation to the time of mating and ovulation. Canad. J. Genet. Cytol. $\underline{11}$:287, 1969.

RODENTIA

Vol. 2, Folio 73

8) Fernández, L.R. and Spotobno, A.: Heteromorphism in chromosome pair No. 1 of Cavia porcellus. Arch. Biol. y Medic. Exper. $\underline{5}$:81, 1969.

9) Jagiello, G.M.: Some cytologic aspects of meiosis in female guinea pig. Chromosoma $\underline{27}$:95, 1969.

RODENTIA

Vol. 1, Folio 13

15) Deaven, L.L. and Stubblefield, E.: Segregation of chromosomal DNA in Chinese hamster fibroblasts in vitro. Exp. Cell Res. 55:132, 1969.

16) Dewey, W.C. and Miller, H.H.: X-ray induction of chromatid exchanges in mitotic and G₁ Chinese hamster cells pretreated with colcemid. Exp. Cell Res. 57:63, 1969.

17) Sonnenschein, C., Roberts, D.W. and Yerganian, G.: Karyotypic and enzymatic characteristics of a somatic hybrid cell line originating from dwarf hamsters. Genetics 62:379, 1969.

18) Yerganian, G., Nell, M.A., Cho, S.S., Hayford, A.H. and Ho, T.: Virus-associated gain and loss of proliferative and neoplastic properties of normal and virus-transformed diploid cell lines. Nat. Cancer Inst. Monogr. No. 29: 241, 1968.

Vol. 1, Folio 14

15) Sasaki, M. and Kamada, T.: A phenotypically normal female golden hamster with sex-chromosome anomaly. Jap. J. Genet. 44:11, 1969.

16) Raicu, P., Ionescu-Varo, M. and Duna, D.: Interspecific crosses between the Rumanian and Syrian hamster. Cytogenetic and histologic studies. J. Hered. 60:149, 1969.

Vol. 1, Folio 17

21) Fraccaro, M., Hansson, K., Hultén, M., Lindsten, J. and Tiepolo, L.: Heterochromatin in preimplantation mouse embryos. Exp. Cell Res. 55:427, 1969.

22) Lyon, M.F.: A true hermaphrodite mouse presumed to be an XO/XY mosaic. Cytogenetics 8:326, 1969.

23) Léonard, A. and Deknudt, Gh.: Etude cytologique d'une translocation chromosome Y-autosome chez la souris. Experientia 25:876, 1969.

24) Léonard, A. and Deknudt, Gh.: Etude d'une translocation de type Robertsonien chez les souris de race AKR. Acta Zool. Path. Antv. 48:43, 1969.

Vol. 2, Folio 71

8) Sen, Y.H.: Karyotypes of Malayan rats (Rodentia-Muridae, genus Rattus Fischer). Chromosoma 27:245, 1969.

RODENTIA

Vol. 1, Folio 18

19) Hori, S.H. and Sasaki, M.: Glucose 6-phosphate dehydrogenase
isoenzyme patterns and chromosomes in primary liver tumors of the rat.
Cancer Res. 29:880, 1969.

20) Sen, Y.H.: Karyotypes of Malayan rats (Rodentia-Muridae, genus Rattus
Fischer). Chromosoma 27:245, 1969.

21) Iversen, J.G.: Phytohemagglutinin response of recirculating and non-
recirculating rat lymphocytes. Exp. Cell Res. 56:219, 1969.

CARNIVORA

Vol. 1, Folio 20

16) Ford, L.: Identification and chromomeric interpretation of pachytene
bivalents from Canis familiaris. Canad. J. Genet. Cytol. 11:389, 1969.

17) Clough, E., Pyle, R.L., Hare, W.C.D., Kelly, D.F. and Patterson, D.F.:
An XXY sex-chromosome constitution in a dog with testicular hypoplasia and
congenital heart disease. Cytogenetics 9:71, 1970.

Vol. 2, Folio 78

3) Benirschke, K.: Zoos and the pathologist - a two way street or
Cytogenetics on zoo animals. Acta Zool. Path. Antv. 48:29, 1969.

PERISSODACTYLA

Vol. 1, Folio 34

9) Pirtle, E.C. and Woods, L.K.: Observations of chromosomes of the horse.
Mammalian Chromosomes Newsletter 9:3, 1968.

ARTIODACTYLA

Vol. 3, Folio 137

5) Koulischer, L.: Concept of cellular clonal evolution of karyotypes
applied to evolution of species. Acta Zool. Path. Antv. 48:21, 1969.

Vol. 3, Folio 138

4) Bhambhani, R. and Kuspira, J.: The somatic karyotypes of American bison
and domestic cattle. Canad. J. Genet. Cytol. 11:243, 1969.

ARTIODACTYLA

Vol. 1, Folio 44

25) Amrud, J.: Centric fusion of chromosomes in Norwegian red cattle (NRF). Hereditas 62:293, 1969.

26) Bhambhani, R. and Kuspira, J.: The somatic karyotypes of American bison and domestic cattle. Canad. J. Genet. Cytol. 11:243, 1969.

27) Gustavsson, J.: Cytogenetics, distribution and phenotypic effects of a translocation in Swedish cattle. Hereditas 63:68, 1969.

Vol. 3, Folio 140

6) Koulischer, L.: Concept of cellular clonal evolution of karyotypes applied to evolution of species. Acta Zool. Path. Antv. 48:21, 1969.

Vol. 1, Folio 45

19) Bruère, A.N.: Male sterility and an autosomal translocation in Romney sheep. Cytogenetics 8:209, 1969.

20) Bruère, A.N., Marshall, R.B. and Ward, D.P.J.: Testicular hypoplasia and XXY sex chromosome complement in two rams: the ovine counterpart of Klinefelter's syndrome in man. J. Reprod. Fert. 19:103, 1969.

21) Koulischer, L.: Concept of cellular clonal evolution of karyotypes applied to evolution of species. Acta Zool. Path. Antv. 48:21, 1969.

22) Jönsson, G. and Gustavsson, I.: Blood cell chimerism in one of three triplet lambs. J. Hered. 60:175, 1969.

Vol. 1, Folio 39

4) Lawrence, H.L.: Wild boars of the Appalachians. Natural History 78:46, 1969.

PRIMATES

Vol. 2, Folio 98

5) Egozcue, J., Perkins, E.M., Hagemenas, F. and Ford, D.M.: The chromosomes of some Platyrrhini (Callicebus, Ateles and Saimiri). Folia primat. 11:17, 1969.

Vol. 1, Folio 48

10) Chiarelli, B.: The phylogeny of primates from a karyological point of view. Acta Zool. Path. Antv. 48:11, 1969.

11) Markarian, D.S., Matchavariani, M.G. and Avjian, M.V.: The normal karyotype of the green monkey (Cercopithecus aethiops). Genetika 5:143, 1969.

Vol. 1, Folio 50

11) Al-Aish, M.S.: Human chromosome morphology. I. Studies on normal chromosome characterization, classification and karyotyping. Canad. J. · Genet. Cytol. 11:370, 1969.

Vol. 3, Folio 147

11) Egozcue, J.: Meiosis in five Macaca species. Folia primat. 11:1, 1969.

Vol. 3, Folio 146

4) Elliot, O.S., Borgaonkar, D.S. and Wong, M.: Chromosome number as related to general biological affinities of tree shrews. Mammalian Chromosomes Newsletter 9:51, 1968.

5) Elliot, O.S., Wong, M. and Borgaonkar, D.S.: Karyological study of Tupaia from Thailand. J. Hered. 60:153, 1969.

Vol. 2, Folio 97

7) Elliot, O.S., Borgaonkar, D.S. and Wong, M.: Chromosome number as related to general biological affinities of tree shrews. Mammalian Chromosomes Newsletter 9:51, 1968.

8) Elliot, O.S., Wong, M. and Borgaonkar, D.S.: Karyological study of Tupaia from Thailand. J. Hered. 60:153, 1969.

Order: MARSUPIALIA

Family: DIDELPHIDAE

Didelphis marsupialis

$2n = 22$

Order: MARSUPIALIA   Family: DIDELPHIDAE

Didelphis marsupialis

2n=22

AUTOSOMES: 20 Acrocentrics

SEX CHROMOSOMES: X Acrocentric
                  Y Acrocentric

    The American opossum, Didelphis marsupialis virginiana Kerr (Folio 51),
possesses a diploid number of 22 but 12 autosomes and the X are biarmed. This
feature is consistent among all specimens collected in USA and in a portion of
Mexico. However, D. marsupialis specimens collected in South America and
Central America displayed no biarmed elements as depicted here. An extensive
study made by Dr. Alfred L. Gardner (personal communication) showed that in
Mexico both forms can be found, but the two are apparently isolated by
ecological barriers. According to Gardner's opinion, the two should be
considered separate species. At any rate, the karyotype in Folio 51 should
have its subspecies name virginiana affixed.

    The karyotypes presented here are gifts of Dr. A. L. Gardner. The male
specimen bears specimen number ALG 11385 and is deposited in the Museum of
Zoology, Louisiana State University, Baton Rouge, Louisiana, USA. The female
specimen was collected 5 Km W. by road from El Naranjo, San Luis Potosí,
Mexico. Bone marrow was used for the male and skin culture was used for
the female specimens.

Order: MARSUPIALIA   Family: DIDELPHIDAE

Didelphis marsupialis

2n=22

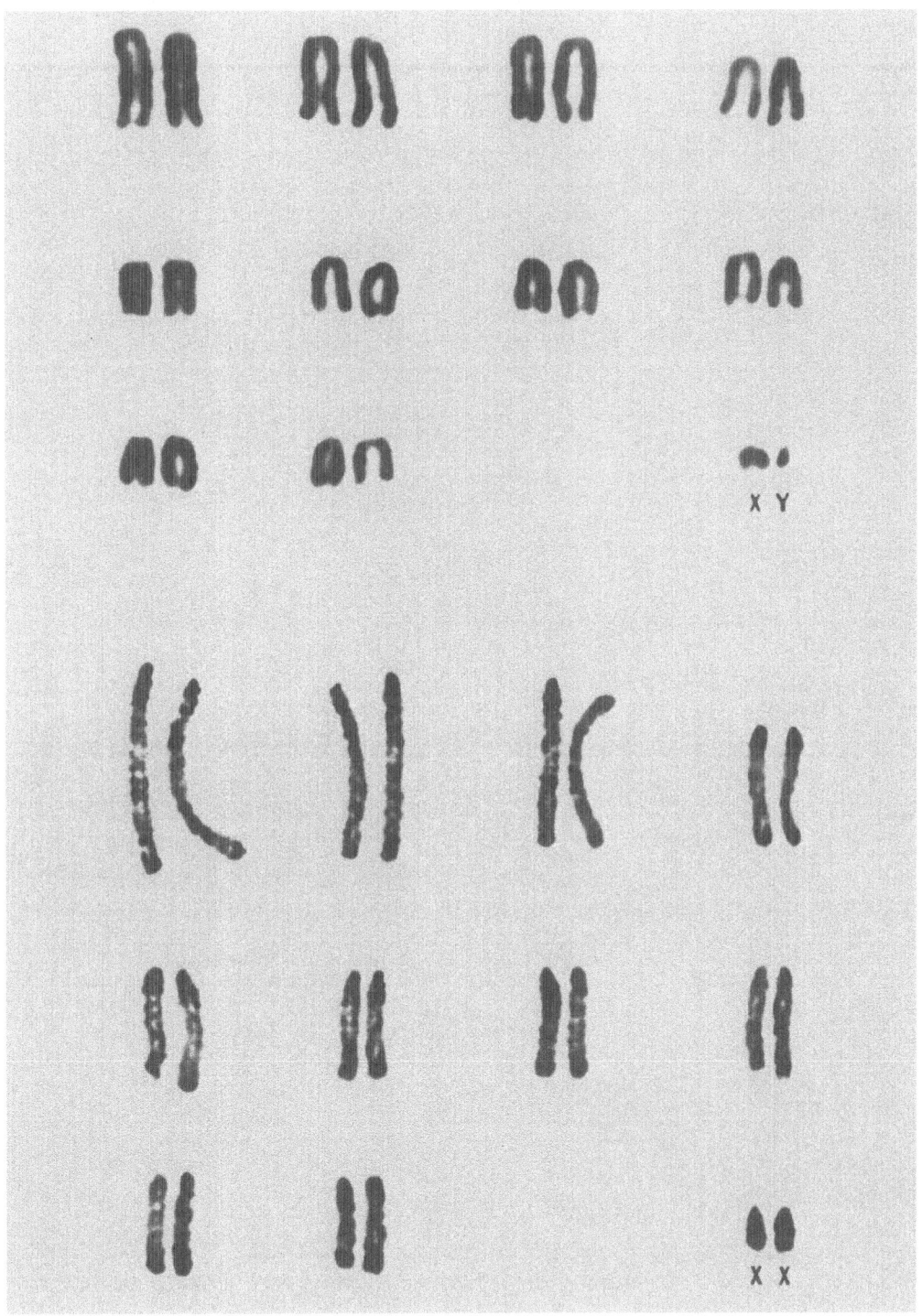

Order: MARSUPIALIA

Family: DIDELPHIDAE

Marmosa alstoni (Alston's opossum)
2n = 14

Volume 5, Folio 202, 1971

Order: MARSUPIALIA   Family: DIDELPHIDAE

<u>Marmosa</u> <u>alstoni</u> (Alston's opossum)

2n=14

AUTOSOMES: 8 Metacentrics and submetacentrics
             2 Subtelocentrics
             2 Acrocentrics

SEX CHROMOSOMES: X Acrocentric
                   Y Acrocentric

The X chromosome can either be called subtelocentric or acrocentric.
It is about as long as, or slightly longer than, the subtelocentric autosome.
This feature can be used to distinguish from the karyotype of <u>M</u>. <u>murina</u>
(Folio 203), whose X is shorter than the subtelocentric autosome.

The karyotypes presented here are gifts of Dr. James L. Patton. The
specimens were originally collected by Dr. Alfred L. Gardner, at ca 2km N.W.
Santa Ana, Prov. San José, Costa Rica. They bear the voucher specimen
numbers ♂12638 and ♀12637, Museum of Zoology, Louisiana State University,
Baton Rouge, Louisiana, USA. Bone marrows were used for cytological
preparations.

Order: MARSUPIALIA

Family: DIDELPHIDAE

<u>Marmosa</u> <u>alstoni</u> (Alston's opossum)

2n=14

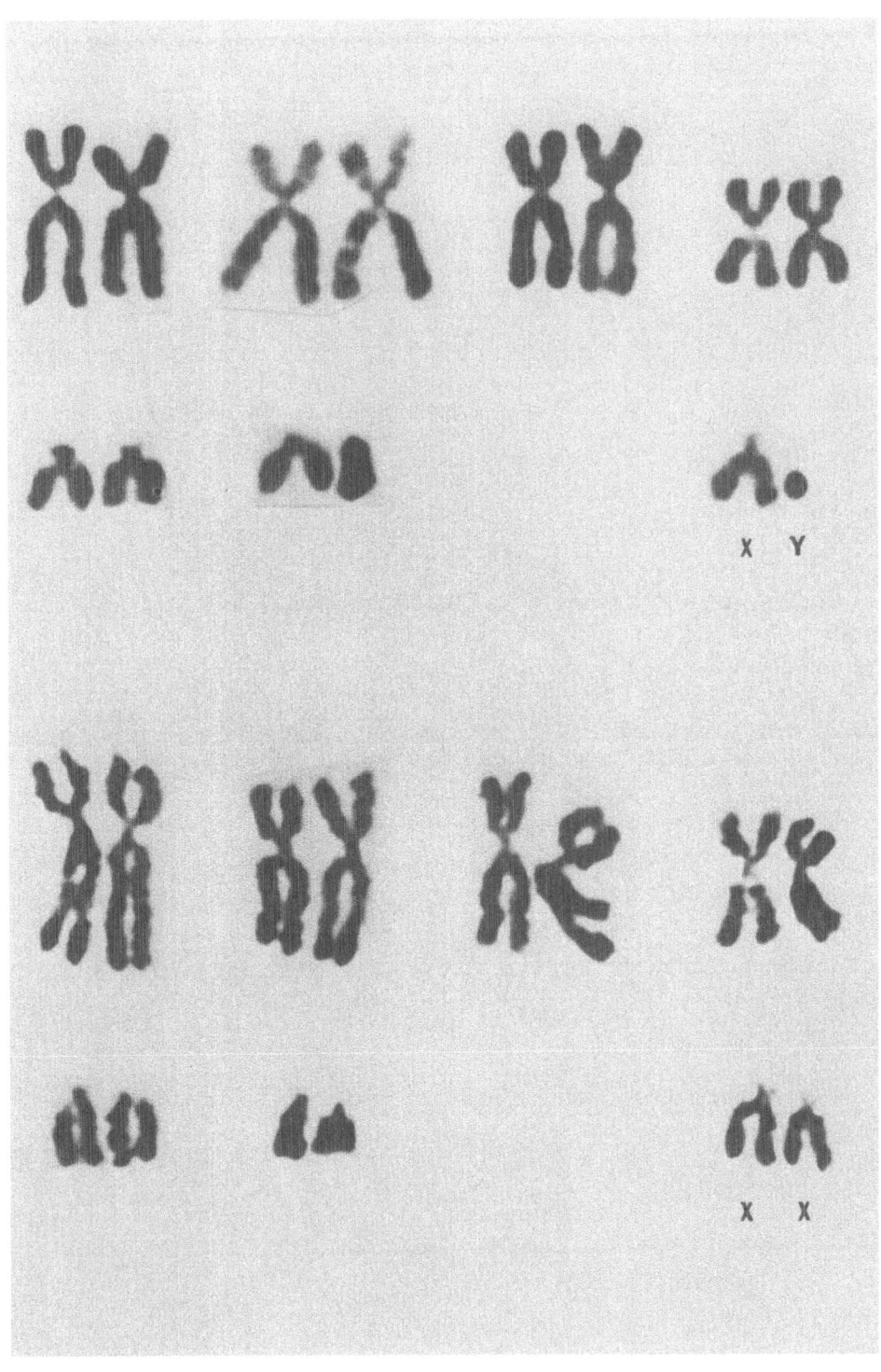

Order: MARSUPIALIA

Family: DIDELPHIDAE

Marmosa murina (Mouse opossum)

$2n = 14$

Order: MARSUPIALIA   Family: DIDELPHIDAE

Marmosa murina (Mouse opossum)

2n=14

AUTOSOMES: 8 Metacentrics and submetacentrics
             2 Subtelocentrics
             2 Acrocentrics

SEX CHROMOSOMES: X Submetacentric
                   Y Acrocentric

The X chromosome may be considered acrocentric. It is the smallest of
the complement besides the Y. This is probably the only feature which can
be distinguished from the karyotype of M. alstoni (Folio 202) whose X
chromosome appears to be as long as the subtelocentric pair. The karyotypes
of Marmosa are also similar to that of Caluromys (Folio 101).

These karyotypes are gifts of Dr. James L. Patton. The specimens were
collected from Balta, Rio Curanja, Depto Loreto, Peru (El. 300 meters). The
specimens are deposited in the Museum of Zoology, University of California at
Berkeley, California, USA, bearing voucher numbers ♂136370 and ♀136368. Bone
marrow was used for cytological preparations.

Marmosa murina (Mouse opossum)

2n=14

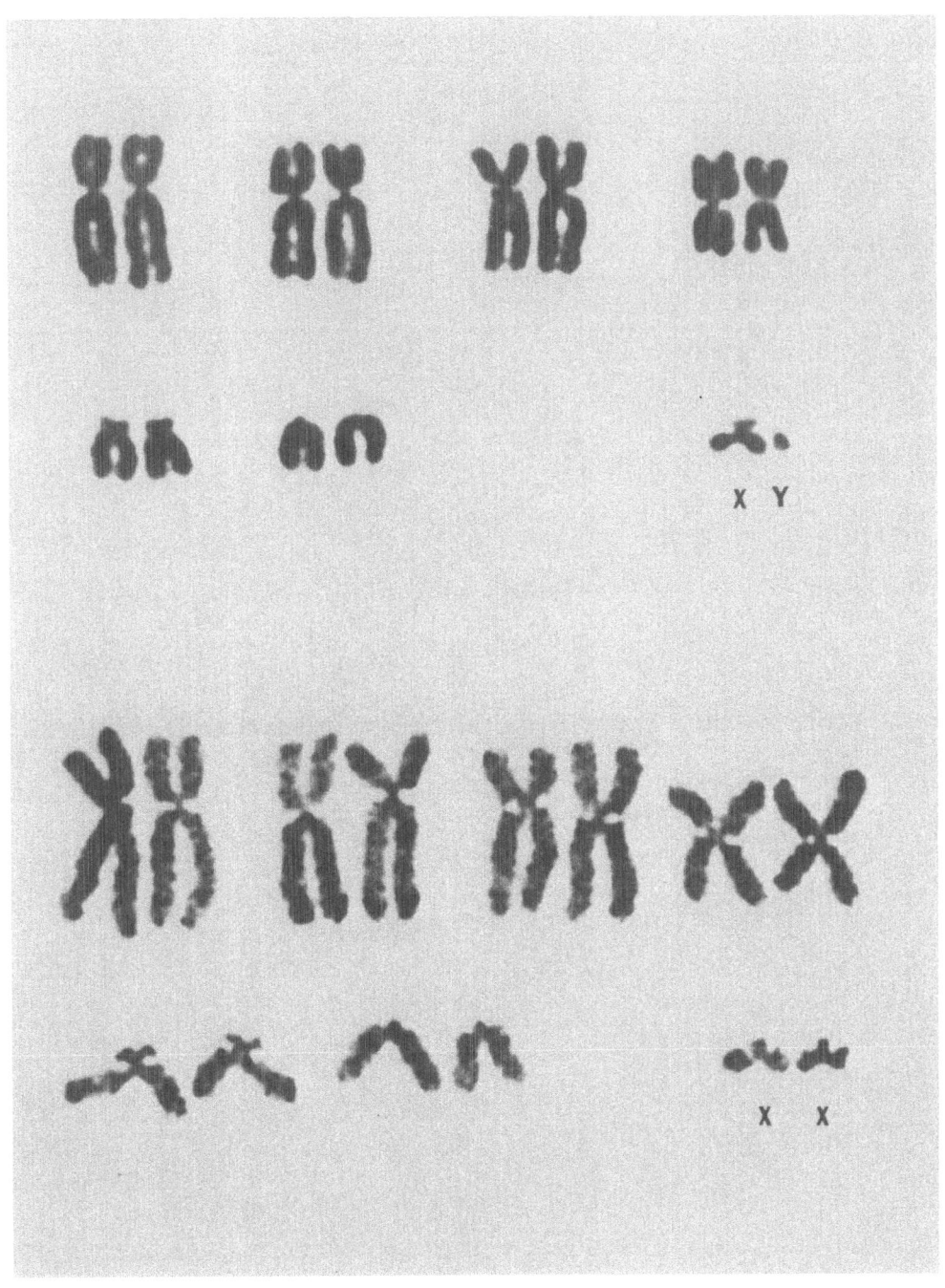

Order: INSECTIVORA

Family: SORICIDAE

Neomys fodiens (Old world water shrew)

$2n = 52$

Order: INSECTIVORA   Family: SORICIDAE

Neomys fodiens (Old world water shrew)

2n=52

AUTOSOMES: 44 Metacentrics, submetacentrics and subtelocentrics
             6 Acrocentrics

SEX CHROMOSOMES: X Submetacentric
                  Y Submetacentric

     The karyotypes were prepared and kindly donated by Dr. K. Fredga,
Lund, Sweden. Five males and two females from Southern Sweden were studied.
Various tissues, including lymphocytes and direct spleen preparations, gave
similar results. The male karyotype displayed came from a lung culture,
that of the female, from a skin culture. Fredga and Levan give a detailed
karyotype description and arm ratios. The X constitutes 5.6% of the haploid
set. Satellite association of many acrocentrics is frequent.

REFERENCES:

1) Bovey, R.: Les chromosomes de Chiroptères et des Insectivores.
Rev. Suisse Zool. 56:371, 1949.

2) Fredga, K. and Levan, A.: The chromosomes of the European water shrew
(Neomys fodiens). Hereditas 62:348, 1969.

Order: INSECTIVORA   Family: SORICIDAE

Neomys fodiens (Old world water shrew)

2n=52

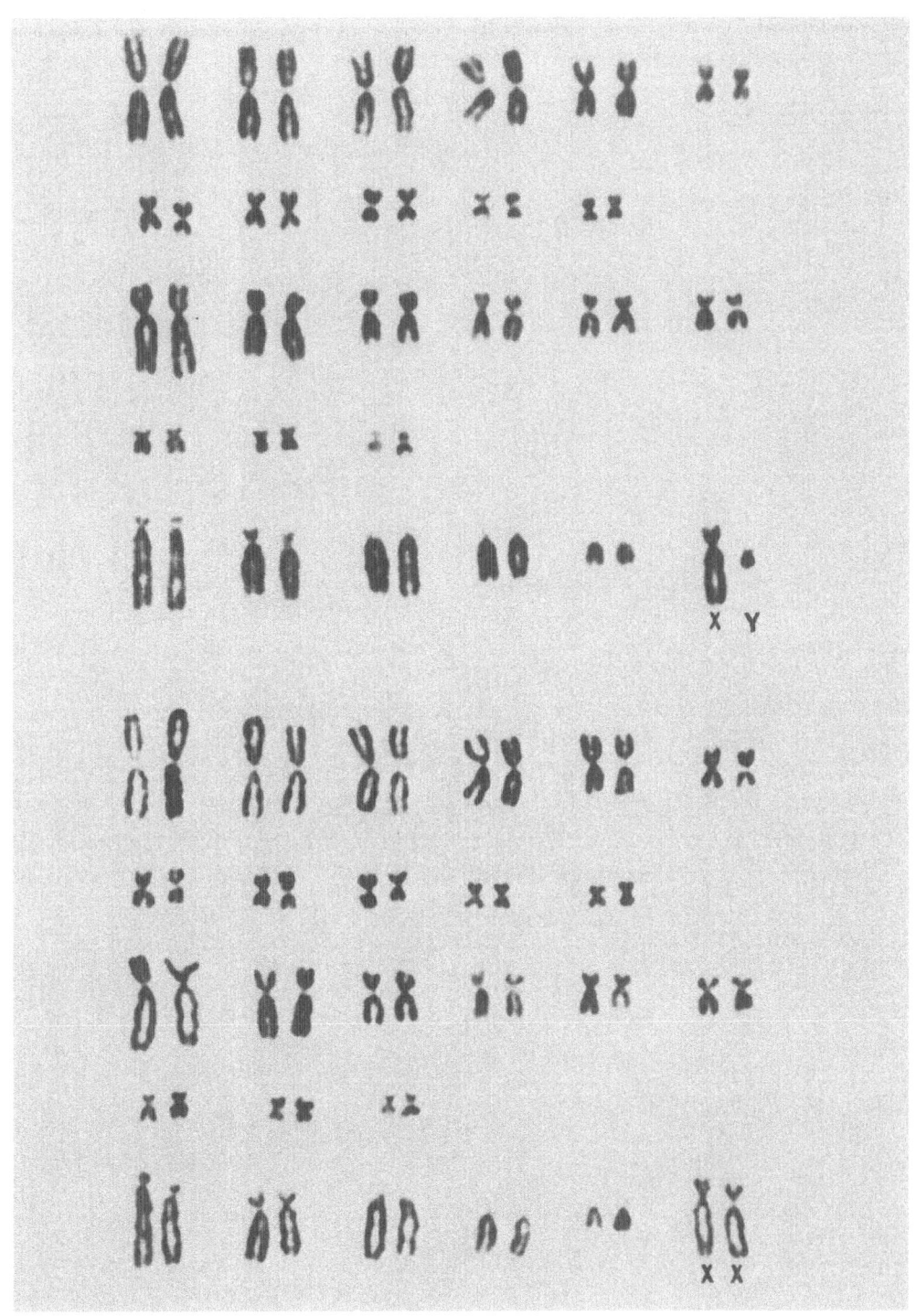

Order: INSECTIVORA

Family: SORICIDAE

Sorex caecutiens (Masked shrew; Laxmann's shrew)
$2n = 42$

Sorex caecutiens (Masked shrew; Laxmann's shrew)

2n=42

AUTOSOMES: 20 Metacentrics and submetacentrics
             20 Acrocentrics

SEX CHROMOSOMES: X Acrocentric
                  Y Acrocentric

    These karyotypes were kindly donated by Dr. K. Fredga, Lund, Sweden, and are arranged according to Fredga's system. Some of the smaller acrocentrics have pronounced short arms and could be considered submetacentrics by others. The smallest autosomes have satellites and engage in satellite association. One male and two females from Sweden were studied and the karyotypes were made from direct spleen preparations.

REFERENCES:

1) Skaren, U. and Halkka, O.: The karyotype of Sorex caecutiens Laxmann. Hereditas 54:376, 1966.

2) Fredga, K.: Chromosomes of the masked shrew (Sorex caecutiens Laxm.) Hereditas 60:269, 1968.

Sorex caecutiens (Masked shrew; Laxmann's shrew)

2n=42

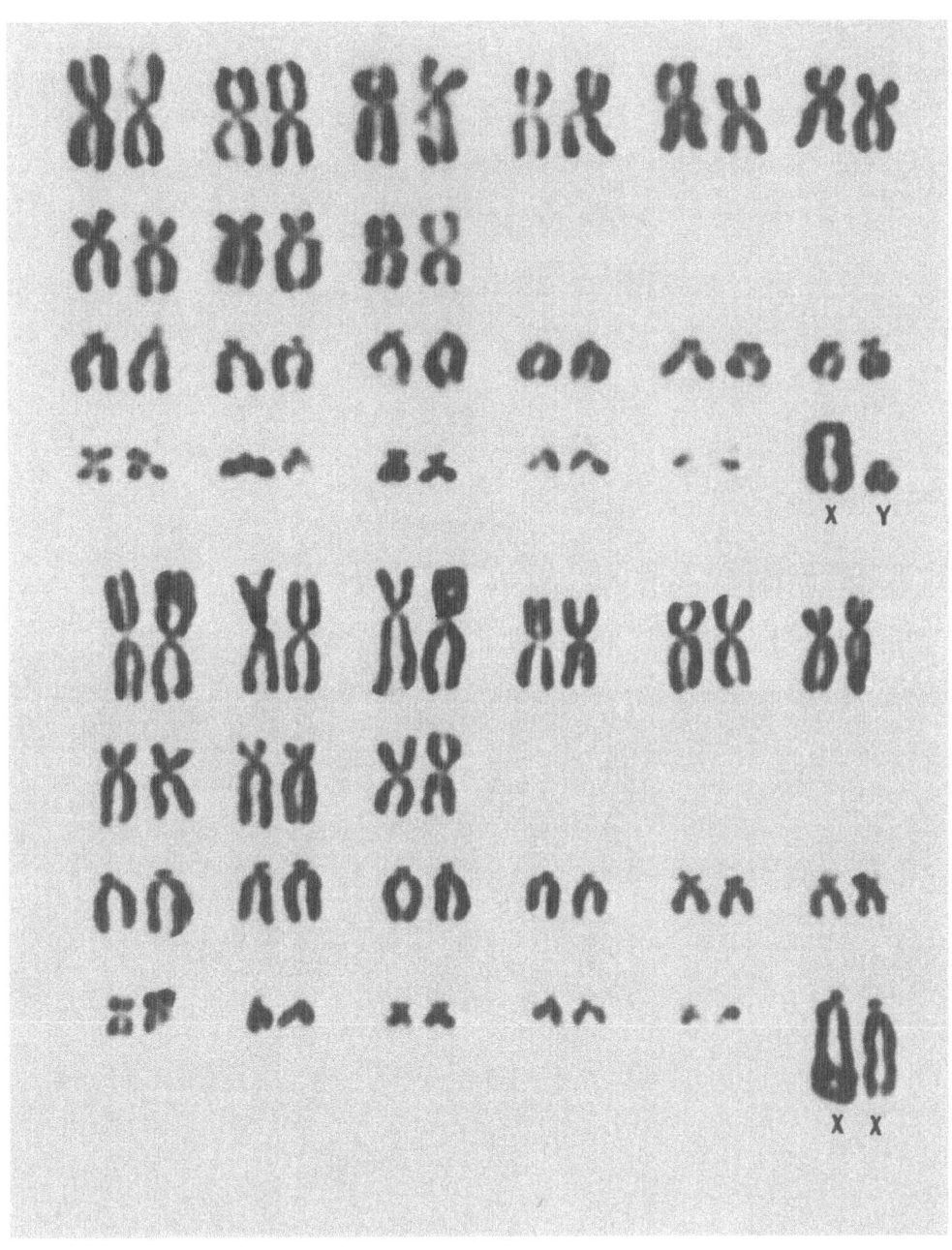

Order: CHIROPTERA

Family: PHYLLOSTOMIDAE

Artibeus turpis (Teapa fruit-eating bat)

$2n = 30$

Order: CHIROPTERA   Family: PHYLLOSTOMIDAE

<u>Artibeus</u> <u>turpis</u> (Teapa fruit-eating bat)

2n=30

AUTOSOMES: 28 Metacentrics and submetacentrics

SEX CHROMOSOMES: X Submetacentric
                   Y Metacentric

     This is one species in the genus <u>Artibeus</u> where a classic XX/XY sex
determination system exists. Thus far the karyotypes of all <u>Artibeus</u> species
(from Central America to the Caribbean Islands) are indistinguishable save for
the Y chromosome constitution.

     The specimens were collected by Dr. Robert J. Baker in Sinaloa, Mexico.
Bone marrows were used for cytological preparations.

REFERENCES:

1) Baker, R.J.: Karyotypes of bats of the family Phyllostomidae and their
taxonomic implications. Southwest. Natural. <u>12</u>:407, 1967.

2) Hsu, T.C., Baker, R.J. and Utakoji, T.: The multiple sex chromosome
system of American leaf-nosed bats (Chiroptera, Phyllostomidae).
Cytogenetics <u>7</u>:27-38, 1968.

Order: CHIROPTERA

Family: PHYLLOSTOMIDAE

Artibeus turpis (Teapa fruit-eating bat)

2n=30

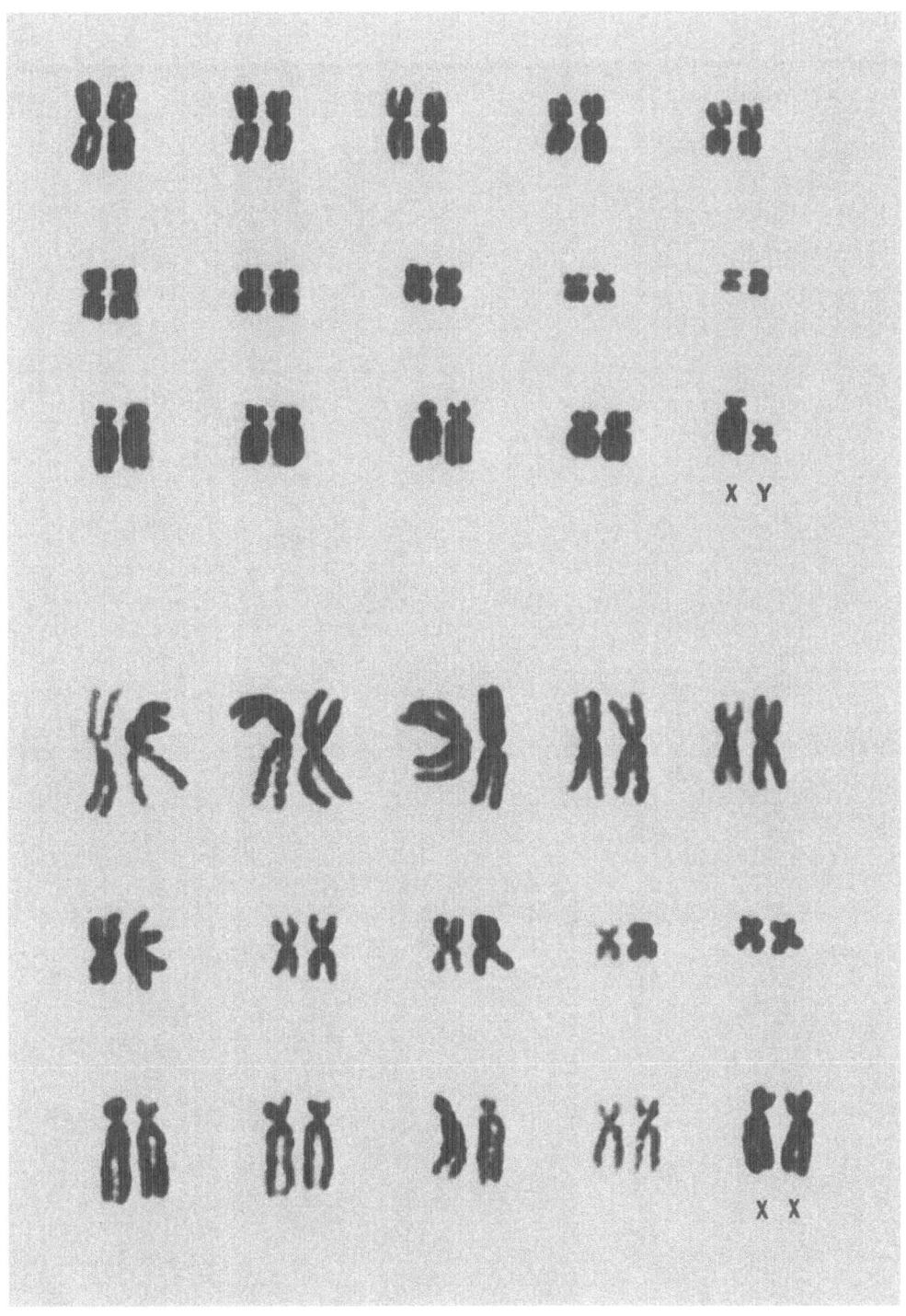

Order: CHIROPTERA

Family: PHYLLOSTOMIDAE

Chiroderma villosum

$2n = 26$

Chiroderma villosum

2n=26

AUTOSOMES: 24 Metacentrics and submetacentrics

SEX CHROMOSOMES: X Chromosome submetacentric
                   Y Chromosome submetacentric

The X chromosomes and three pairs of autosomes have distinctly unequal arm ratios. Identification of the X is equivocal. The Y is the smallest element of the complement.

The specimens were collected by Dr. Robert J. Baker from San Fernando Ranch, Chiapas, Mexico. Bone marrows were used for cytological preparations. The specimen number for the male is 34-109172, and the female, 34-109193. Both are in the collection of the Texas Technological University, Lubbock, Texas, USA.

REFERENCES:

1) Baker, R.J.: Karyotypes of bats of the family Phyllostomidae and their taxonomic implications. Southwest. Natural. 12:407, 1967.

2) Hsu, T.C., Baker, R.J. and Utakoji, T.: The multiple sex chromosome system of American leaf-nosed bats (Chiroptera, Phyllostomidae). Cytogenetics 7:27-38, 1968.

Chiroderma villosum

2n=26

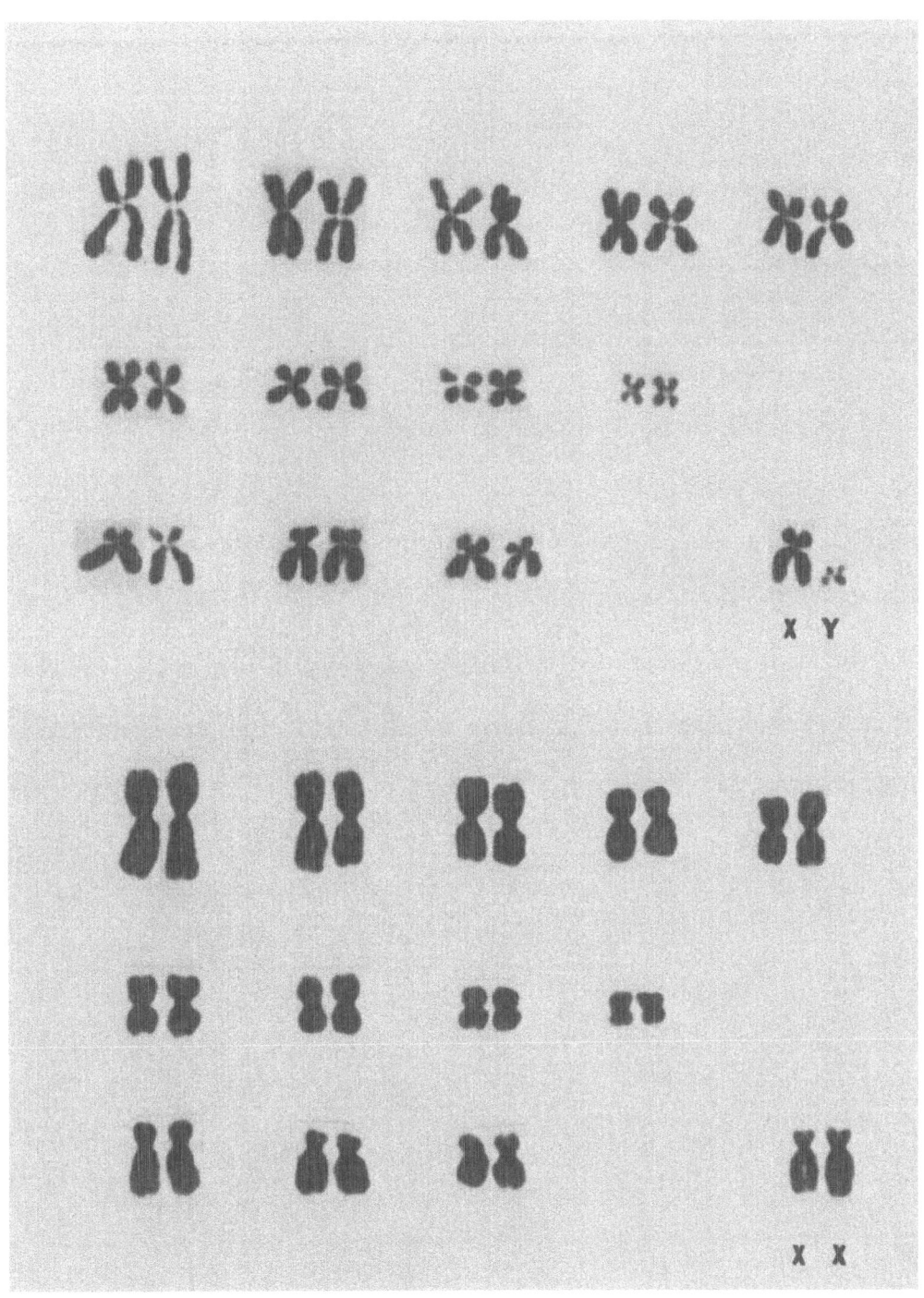

Order: CHIROPTERA

Family: PHYLLOSTOMIDAE

Glossophaga soricina (Pallas' long-tongued bat)

$2n = 32$

Glossophaga soricina (Pallas' long-tongued bat)

2n=32

AUTOSOMES: 30 Metacentrics and submetacentrics

SEX CHROMOSOMES: X Submetacentric
                  Y Minute acrocentric

    According to Baker (1967), all the species in the genus Glossophaga studied thus far, G. soricina (Pallas), G. alticola Davis, and G. commissarisi Gardner have indistinguishable karyotypes. The smallest autosomal pair bears a secondary constriction on one arm near the centromere. Identification of the X is equivocal, but identification of the Y is absolute.

    The karyotypes presented here came from specimens collected by Dr. Robert J. Baker. The male specimen was caught 231 Km from Mexico City by Highway 95 to Acapulco, Guerrero, Mexico, and the female, 157 Km by Highway 95 south of Mexico City, Guerrero, Mexico. Bone marrows were used for cytological preparations.

REFERENCES:

1) Baker, R.J.: Karyotypes of bats of the family Phyllostomidae and their taxonomic implications. Southwestern Natural. 12:407, 1967.

2) Hsu, T.C., Baker, R.J. and Utakoji, T.: The multiple sex chromosome system of American leaf-nosed bats (Chiroptera, Phyllostomidae). Cytogenetics 7:27, 1968.

Order: CHIROPTERA

Family: PHYLLOSTOMIDAE

Glossophaga soricina (Pallas' long-tongued bat)

2n=32

Order: CHIROPTERA

Family: PHYLLOSTOMIDAE

Pteronotus parnellii

$2n = 38$

Pteronotus parnellii

2n=38

AUTOSOMES: 24 Metacentrics and submetacentrics
             12 Acrocentrics

SEX CHROMOSOMES: X Submetacentric
                   Y Acrocentric

According to Baker (1967), three species of this genus, Pt. parnellii (Gray), Pt. psilotus (Dobson), and Pt. davyi (Gray), possess identical karyotypes. The identification of the X chromosome is somewhat subjective, but it is one of the medium-sized submetacentrics with nearly equal arms. The Y chromosome knob-like second arm of the Y chromosome may be obscure in some cells.

The specimens were collected by Dr. Robert J. Baker. The male was caught 231 Km from Mexico City by Highway 95 to Acapulco, Guerrero, Mexico, and the female, from Km 184 on Highway 200 N. of Huixulta, Chiapas, Mexico. Bone marrows were used for cytological preparations.

REFERENCES:

1) Baker, R.J.: Karyotypes of bats of the family Phyllostomidae and their taxonomic implications. Southwestern Natural. 12:407, 1967.

2) Hsu, T.C., Baker, R.J. and Utakoji, T.: The multiple sex chromosome system of American leaf-nosed bats (Chiroptera, Phyllostomidae). Cytogenetics 7:27, 1968.

Pteronotus parnellii

2n=38

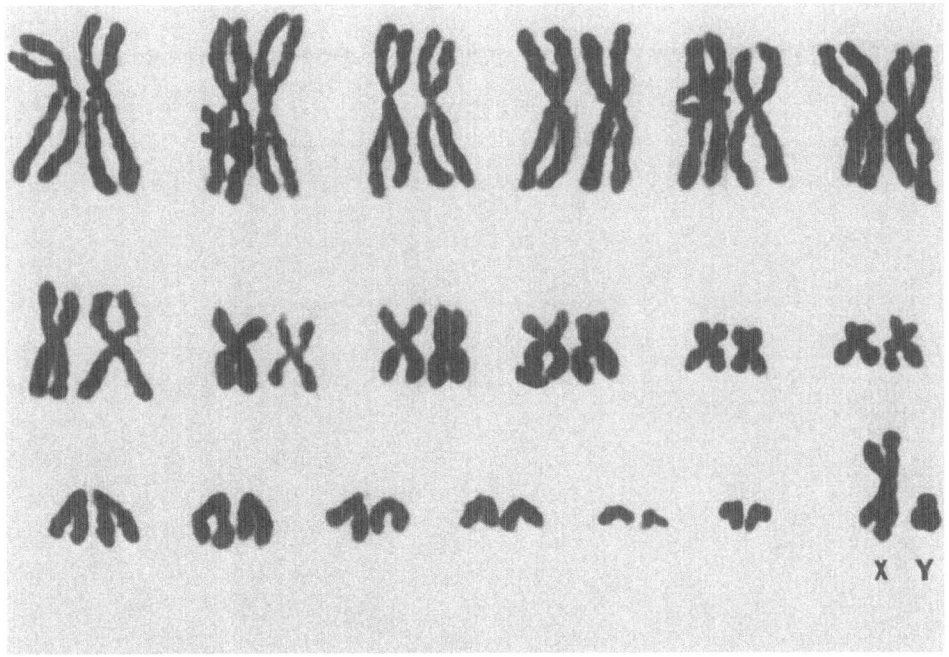

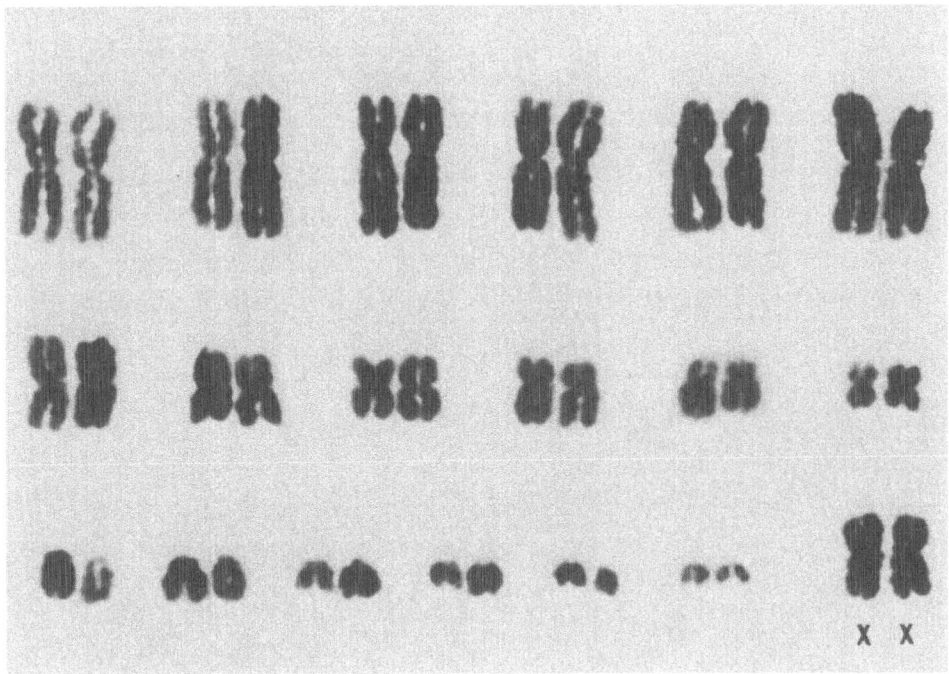

Order: CHIROPTERA

Family: VESPERTILIONIDAE

Nycticeius humeralis (Evening bat)

$2n = 46$

Order: CHIROPTERA   Family: VESPERTILIONIDAE

<u>Nycticeius</u> <u>humeralis</u> (Evening bat)

2n=46

AUTOSOMES: 4 Metacentrics
             40 Acrocentrics

SEX CHROMOSOMES: X Submetacentric
                   Y Minute

     One pair of the medium-sized acrocentrics has a secondary constriction
near the centromere. The two pairs of large, metacentric autosomes are
indistinguishable, but they both are considerably longer than the X.
Identification of the X chromosome is therefore unequivocal. Although
several pairs of autosomes are very small, the Y is the smallest element
of the complement.

     The specimens were collected in the vicinity of Houston, Texas, USA.
Lung cultures were used for cytological preparations.

REFERENCES:

1) Baker, R.J. and Patton, J.L.: Karyotypes and karyotypic variation of
North American Vespertilionid bats. J. Mammal. <u>48</u>:270, 1967.

Order: CHIROPTERA   Family: VESPERTILIONIDAE

Nycticeius humeralis (Evening bat)

2n=46

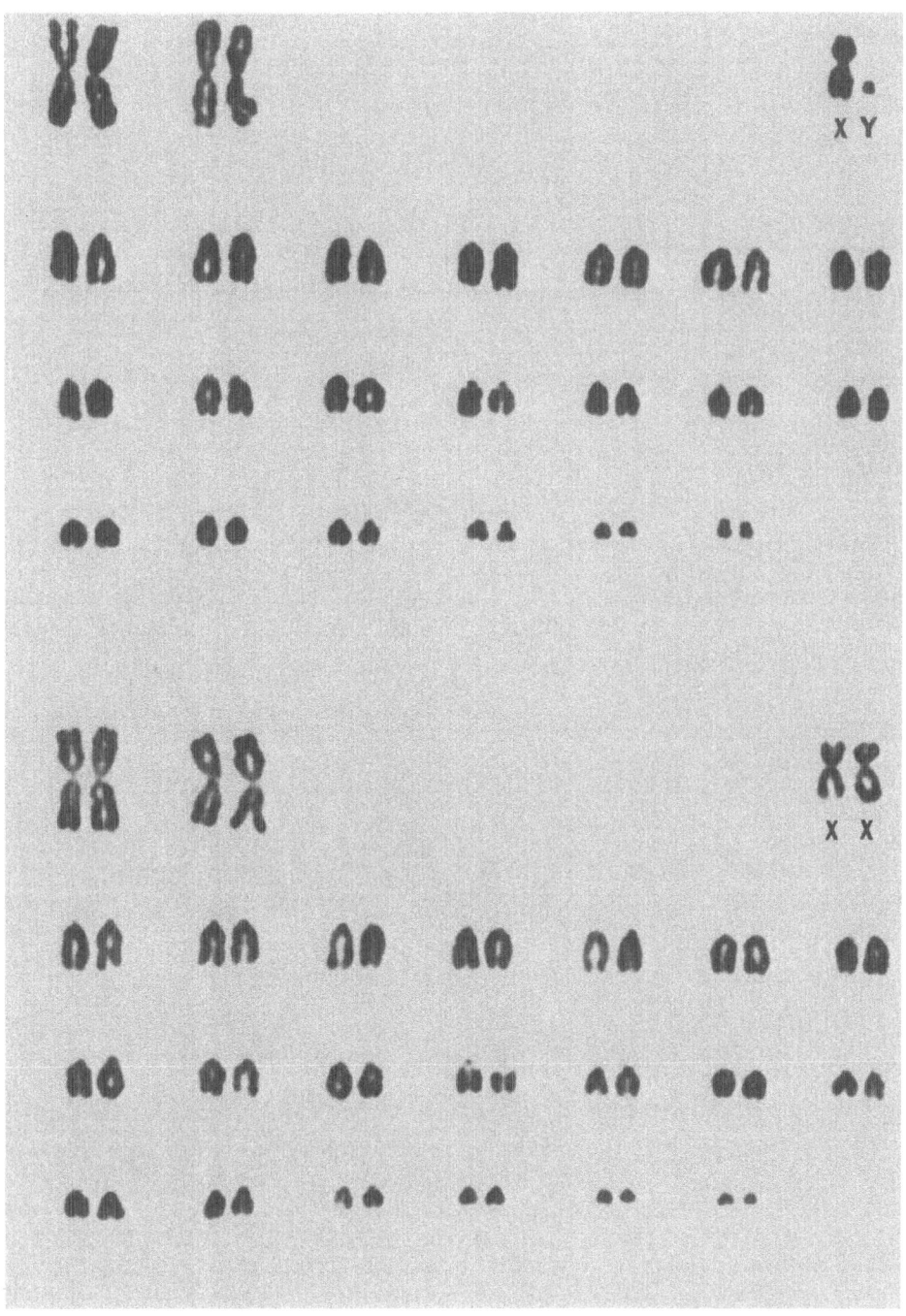

Order: LAGOMORPHA

Family: LEPORIDAE

Lepus townsendii (White-tailed jack rabbit)

$2n = 48$

<u>Lepus townsendii</u> (White-tailed jack rabbit)

2n=48

AUTOSOMES: 42 Metacentrics and submetacentrics
            4 Acrocentrics

SEX CHROMOSOMES: X Submetacentric
                  Y Acrocentric

    Skin biopsies of two animals trapped at Ferguson County, Montana, USA, were kindly made available by Dr. B. W. O'Gara, Missoula, Montana. As with the two previously displayed jack rabbits (Folios 6, 7) the short arms of many autosomes differ in size and classification as to acrocentrics or subtelocentrics is highly arbitrary. These three species have very similar, if not identical, karyotypes.

REFERENCES:

1) Jalal, S.M., James, T.R. and Seabloom, R.W.: Karyotype of white-tailed jack rabbit. Mammalian Chromosomes Newsletter <u>8</u>:194, 1967.

Lepus townsendii (White-tailed jack rabbit)

2n=48

Order: LAGOMORPHA

Family: LEPORIDAE

Sylvilagus floridanus (Eastern cottontail)

$2n = 42$

Sylvilagus floridanus (Eastern cottontail)

2n=42

AUTOSOMES: 40 Metacentrics, submetacentrics and subtelocentrics

SEX CHROMOSOMES: X Submetacentric
                  Y Acrocentric

At least four pairs of subtelocentrics bear satellites on the short arm. Identification of the X chromosome is ambiguous by morphology alone, but it is one of the medium-sized submetacentrics. The Y is the smallest element of the complement.

The male specimen was obtained in Houston, Texas, USA. Lung cultures were initiated for cytological preparations. The female karyotype presented here is a gift of Dr. Katherine G. Palmer.

REFERENCES:

1) Palmer, C. and Armstrong, R.: Chromosome number and karyotype of Sylvilagus floridanus, the Eastern cottontail. Mammalian Chromosomes Newsletter 8:282, 1967.

2) Holden, H.E.: The karyotypes of two species of Sylvilagus. Mammalian Chromosomes Newsletter 9:230, 1968.

Order: LAGOMORPHA   Family: LEPORIDAE

Sylvilagus floridanus (Eastern cottontail)

2n=42

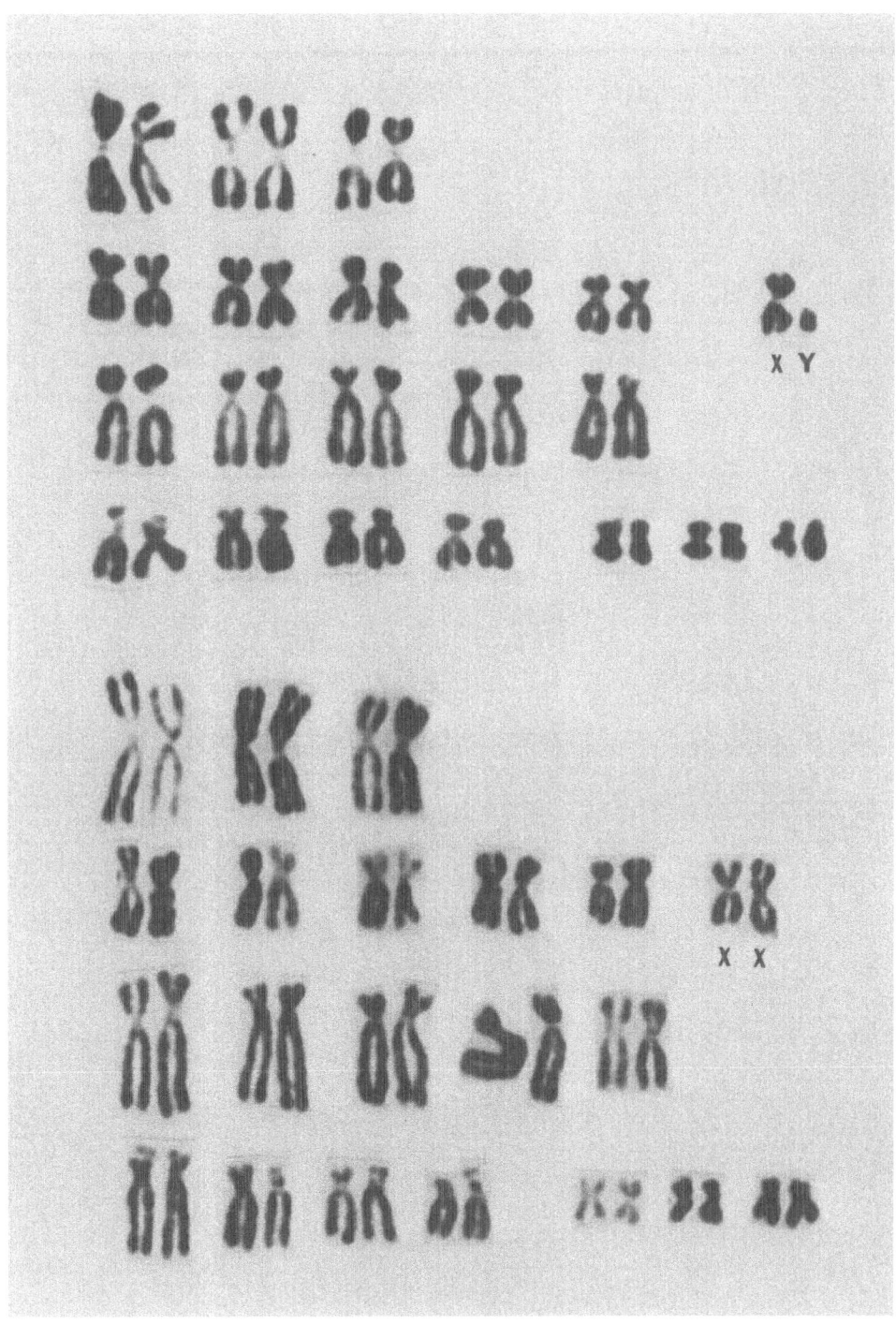

Order: RODENTIA

Family: SCIURIDAE

Ammospermophilus harrisii (Harris' antelope squirrel)

2n = 32

<u>Ammospermophilus</u> <u>harrisii</u> (Harris' antelope squirrel)

2n=32

AUTOSOMES: 30 Metacentrics and submetacentrics

SEX CHROMOSOMES: X Submetacentric
                 Y Acrocentric

The X chromosome is one of the large submetacentric elements, but its identification is not unequivocal. The Y chromosome is the only acrocentric of the complement.

The specimens were collected 1 mile north of Kingman, Mohave County, Arizona, USA. Bone marrows were used for cytological preparations. The specimens are stored in Texas Technological University, Lubbock, Texas, USA, bearing specimen numbers ♀ TCH 1248, ♂ TCH 1249.

REFERENCES:

1) Nadler, C.F. and Sutton, D.A.: Mitotic chromosomes of some North American Sciuridae. Proc. Soc. Exp. Biol. & Med. <u>110</u>:36, 1962.

Order: RODENTIA

Family: SCIURIDAE

Ammospermophilus harrisii (Harris' antelope squirrel)

2n=32

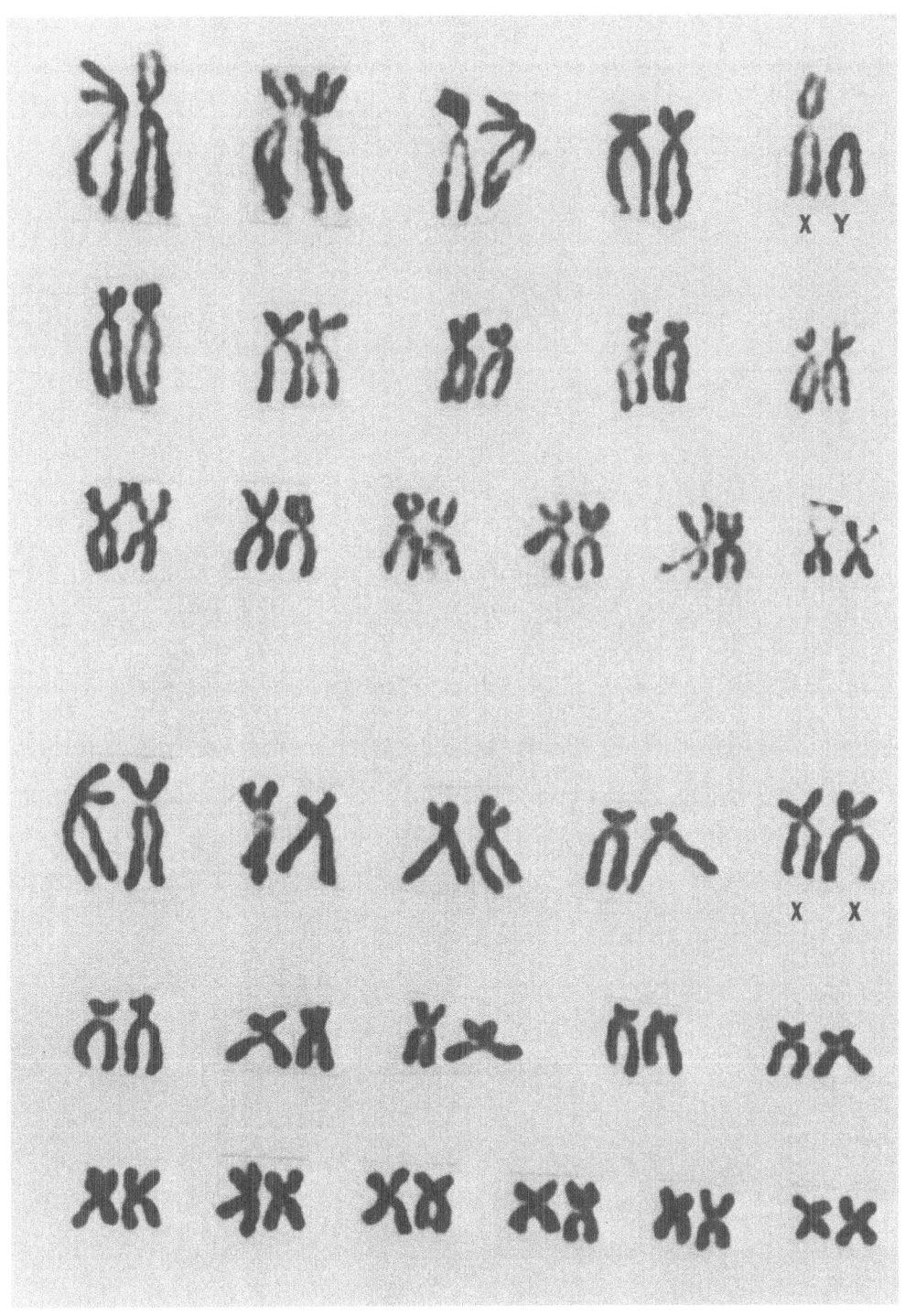

Order: RODENTIA

Family: SCIURIDAE

Callospermophilus (Citellus) lateralis
(Golden-mantled ground squirrel)

$2n = 42$

Order: RODENTIA   Family: SCIURIDAE

Callospermophilus (Citellus) lateralis
(Golden-mantled ground squirrel)

2n=42

AUTOSOMES: 40 Metacentrics and submetacentrics

SEX CHROMOSOMES: X Submetacentric
                  Y Acrocentric

   Skin biopsies from two animals trapped in Missoula County, Montana,
USA, were kindly made available by Dr. B. W. O'Gara.

REFERENCES:

1) Nadler, C.F. and Sutton, D.A.: Mitotic chromosomes of some North
American Sciuridae. Proc. Soc. Exp. Biol. Med. 110:36, 1962.

2) Nadler, C.F.: Chromosomes of Spermophilus franklini and taxonomy of
the ground squirrel genus Spermophilus. Syst. Zool. 15:199, 1966.

Callospermophilus (Citellus) lateralis
(Golden-mantled ground squirrel)

2n=42

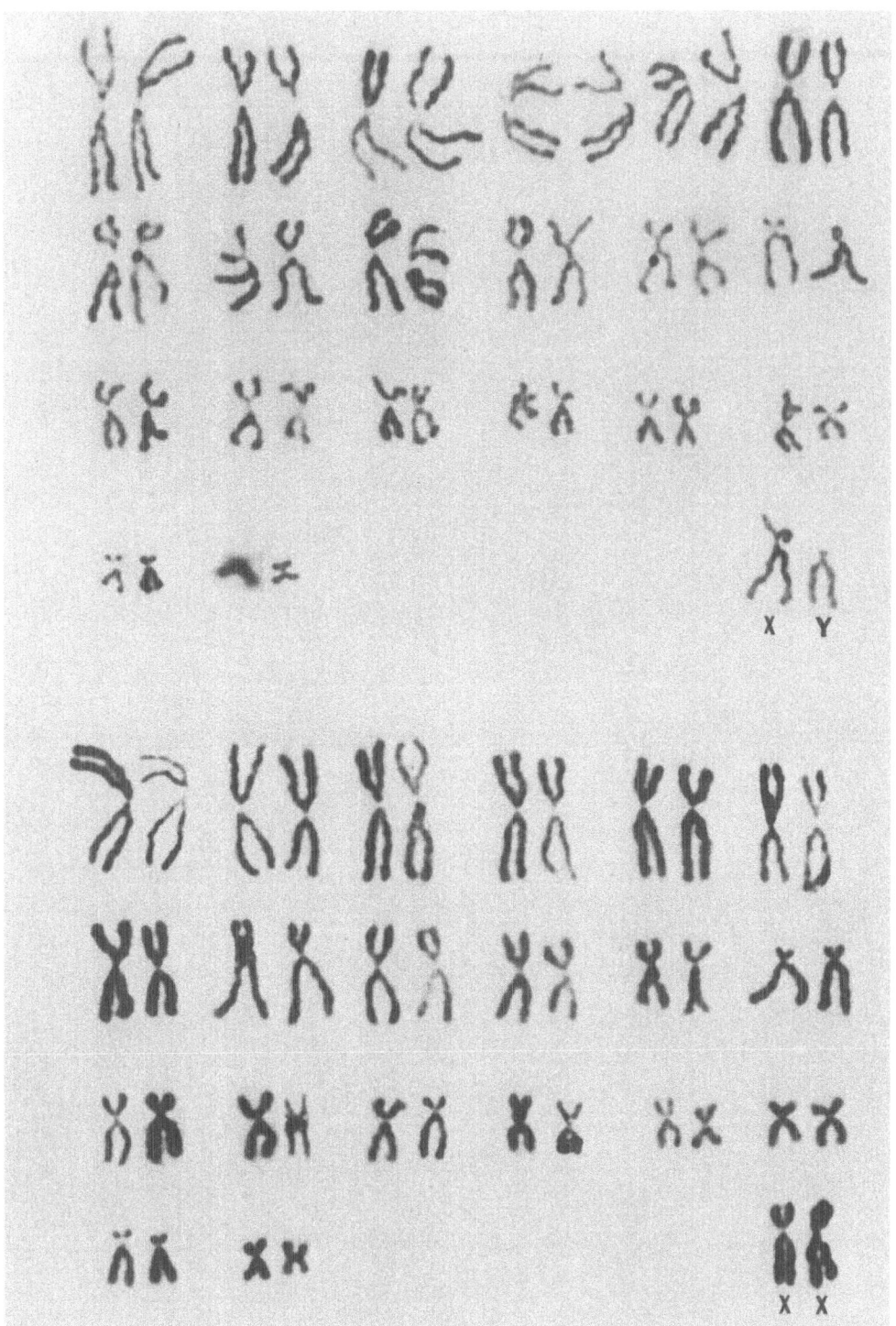

Order: RODENTIA

Family: SCIURIDAE

Cynomys gunnisoni (Gunnison's Prairie dog)
$2n = 40$

Order: RODENTIA   Family: SCIURIDAE

Cynomys gunnisoni (Gunnison's Prairie dog)

2n=40

AUTOSOMES: 26 Metacentrics and submetacentrics
             12 Subtelocentrics or acrocentrics

SEX CHROMOSOMES: X Submetacentric
                  Y Acrocentric

     Most of the "subtelocentrics or acrocentrics" have distinct short arms.
The X chromosome is among the large submetacentrics, and the Y, the smallest
acrocentric.

     The karyotypes presented here are gifts of Dr. Charles F. Nadler,
Northwestern University Medical School, Chicago, Illinois, USA. The
specimens belong to the subspecies C. g. zuniensis Hollister, and were
collected from Flagstaff, Coconino County, Arizona, USA, by Dr. Terry A.
Vaughan. Bone marrows were used for cytological preparations.

REFERENCES:

1) Nadler, C.F. and Hoffmann, R.S.: Chromosomes and serum proteins of some
prairie dogs (Cynomys). J. Mammal. (in press)

Cynomys gunnisoni (Gunnison's Prairie dog)

2n=40

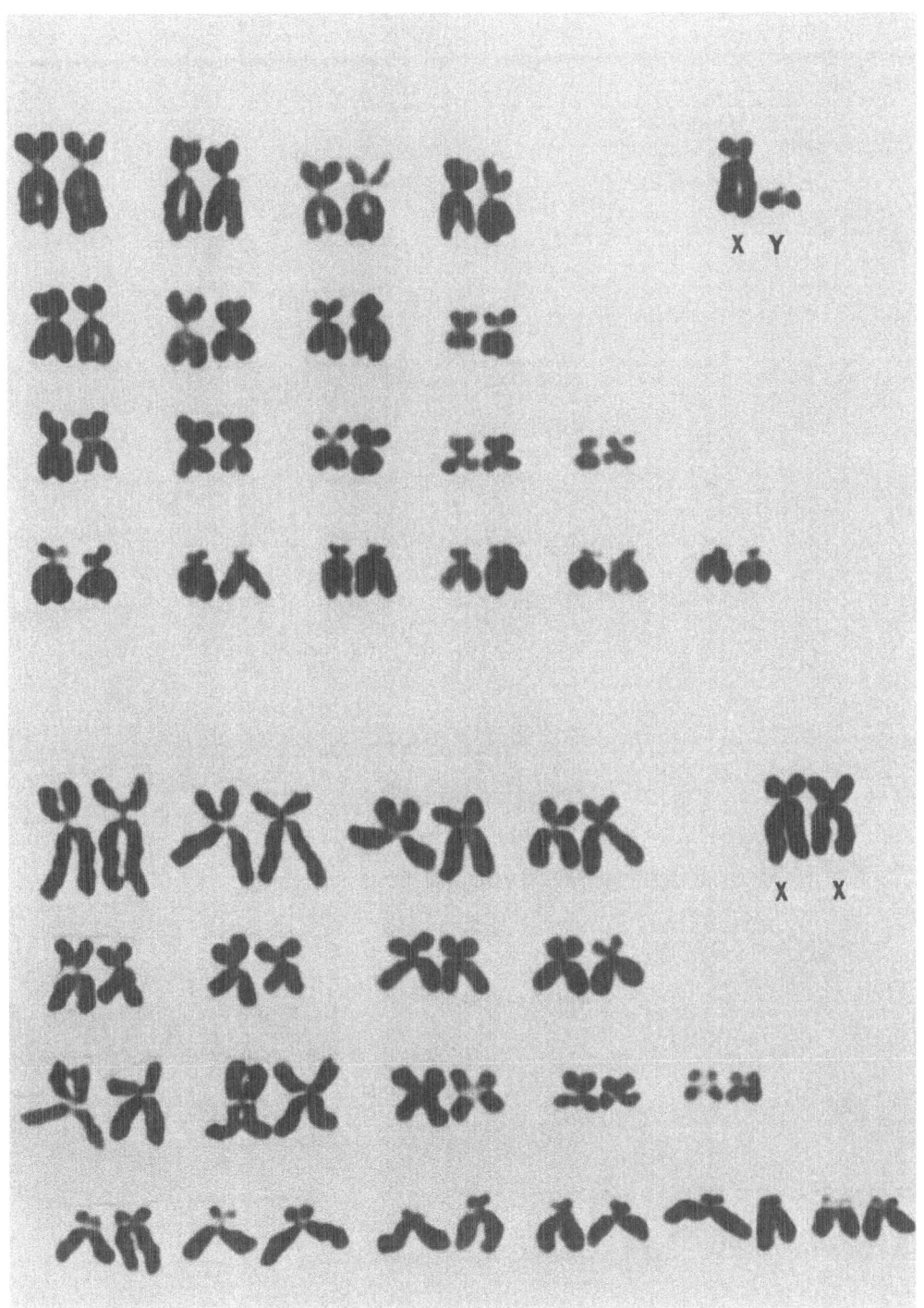

Order: RODENTIA

Family: SCIURIDAE

Eutamias ruficaudus (Red-tailed chipmunk)

$2n = 38$

Order: RODENTIA   Family: SCIURIDAE

<u>Eutamias</u> <u>ruficaudus</u> (Red-tailed chipmunk)

2n=38

AUTOSOMES: 22 Metacentrics and submetacentrics
             14 Acrocentrics

SEX CHROMOSOMES: X Submetacentric
                   Y Metacentric

    Lung, skin and spleen biopsies of two male and two female animals
trapped in Missoula County, Montana, USA, were kindly made available by
Dr. B. W. O'Gara. The karyotype of the male came from a spleen culture,
and that of the female from a skin culture. At least one of the acrocentric
elements displayed here has distinct but small short arm. Other
investigators may prefer to consider this a subtelocentric or even a
submetacentric element.

REFERENCES:

1) Nadler, C. F.: Contributions of chromosomal analysis to the
systematics of North American chipmunks. Amer. Midland Nat. <u>72</u>:298, 1964.

2) Sutton, D. A. and Nadler, C. F.: Chromosomes of the North American
chipmunk genus <u>Eutamias</u>. J. Mammal. <u>50</u>:524, 1969.

Eutamias ruficaudus (Red-tailed chipmunk)

2n=38

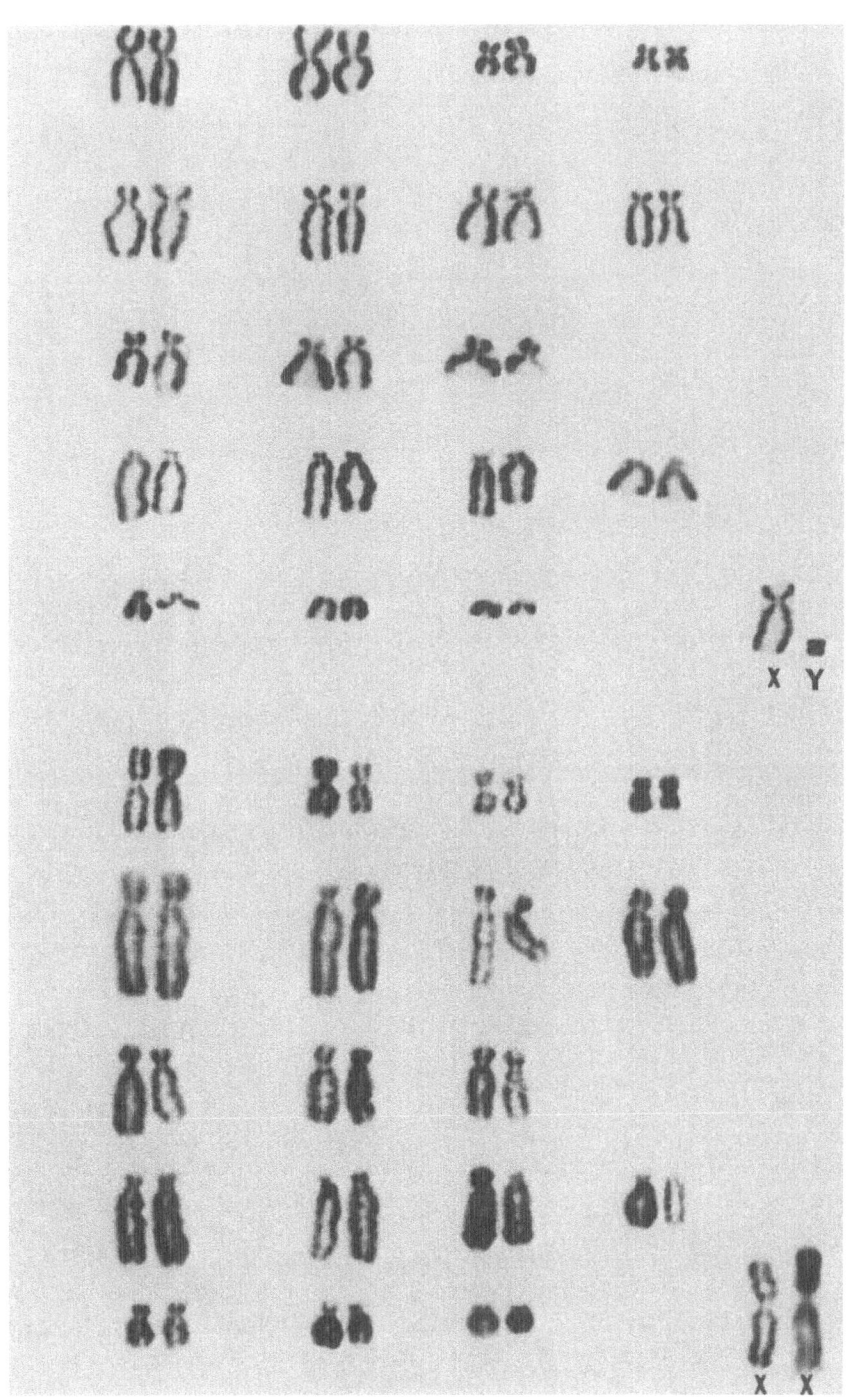

Order: RODENTIA

Family: SCIURIDAE

Eutamias sibiricus (Asian chipmunk)
$2n = 38$

Eutamias sibiricus (Asian chipmunk)

2n=38

AUTOSOMES: 20 Metacentrics, submetacentrics and subtelocentrics
             16 Acrocentrics

SEX CHROMOSOMES: X Submetacentric
                  Y Metacentric

        Several pairs of autosomes can be identified without great difficulty.
The acrocentrics can be roughly classified into two groups: A group of three
pairs of larger elements and another of five pairs of smaller elements. The
X chromosome, however, resembles some of the autosomes.

        The karyotypes are gifts of Dr. Charles F. Nadler. Another set of
karyotypes donated by Dr. M. Sasaki showed similar results.

REFERENCES:

1) Nadler, C.F., Hoffmann, R.S. and Lay, D.M.: Chromosomes of the Asian
chipmunk Eutamias sibiricus Laxmann (Rodentia, Sciuridae). Experientia 25:868,
1969.

2) Sasaki, M., Shimba, H. and Itoh, M.: Karyotypes of two species of
Asiatic squirrels. Mammalian Chromosomes Newsletter 10:227, 1969.

Eutamias sibiricus (Asian chipmunk)

2n=38

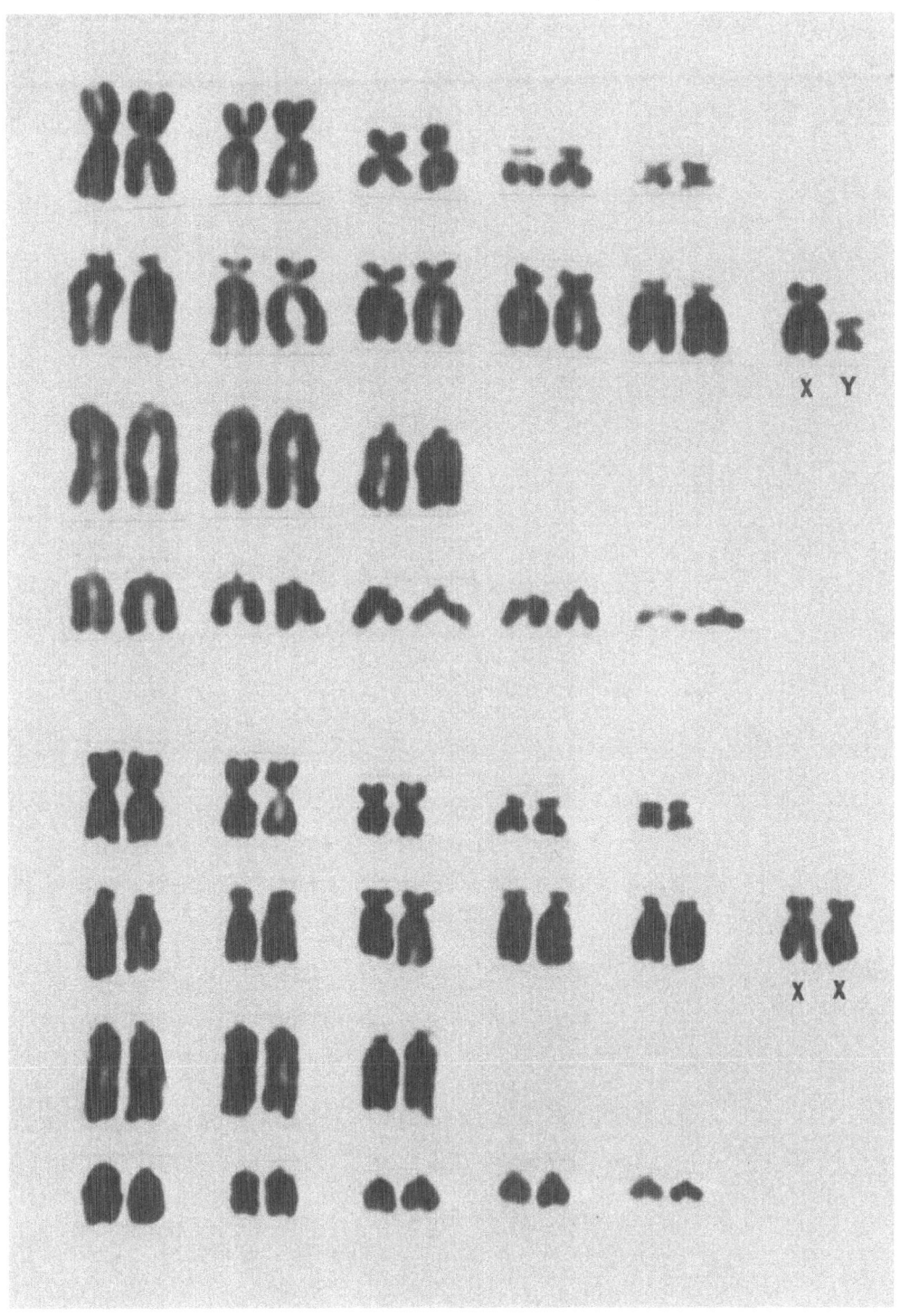

Order: RODENTIA

Family: CRICETIDAE

Peromyscus floridanus (Florida mouse)

$2n = 48$

Order: RODENTIA   Family: CRICETIDAE

<u>Peromyscus floridanus</u> (Florida mouse)

2n=48

AUTOSOMES: 16 Submetacentrics
             30 Acrocentrics

SEX CHROMOSOMES: X Submetacentric with highly disproportioned arms
                   Y Small submetacentric

    These specimens were collected in the vicinity of Sumner, Florida, USA,
in 1967. Samples from Delray Beach, Florida, USA, collected by Dr. J.A. King
showed identical karyotypes. All cytological preparations were made from
primary lung cultures.

    The B-group chromosomes (see Hsu and Arrighi, 1968), are placed in the
last of the karyotypes presented here. The X chromosome appears to be the
second longest submetacentric of the complement. Identification of the Y
offers no great difficulty.

REFERENCES:

1) Hsu, T.C. and Arrighi, F.E.: Chromosomes of <u>Peromyscus</u> (Rodentia,
Cricetidae). I. Evolutionary trends in 20 species. Cytogenetics <u>7</u>:417,
1968.

Order: RODENTIA

Peromyscus floridanus (Florida mouse)

2n=48

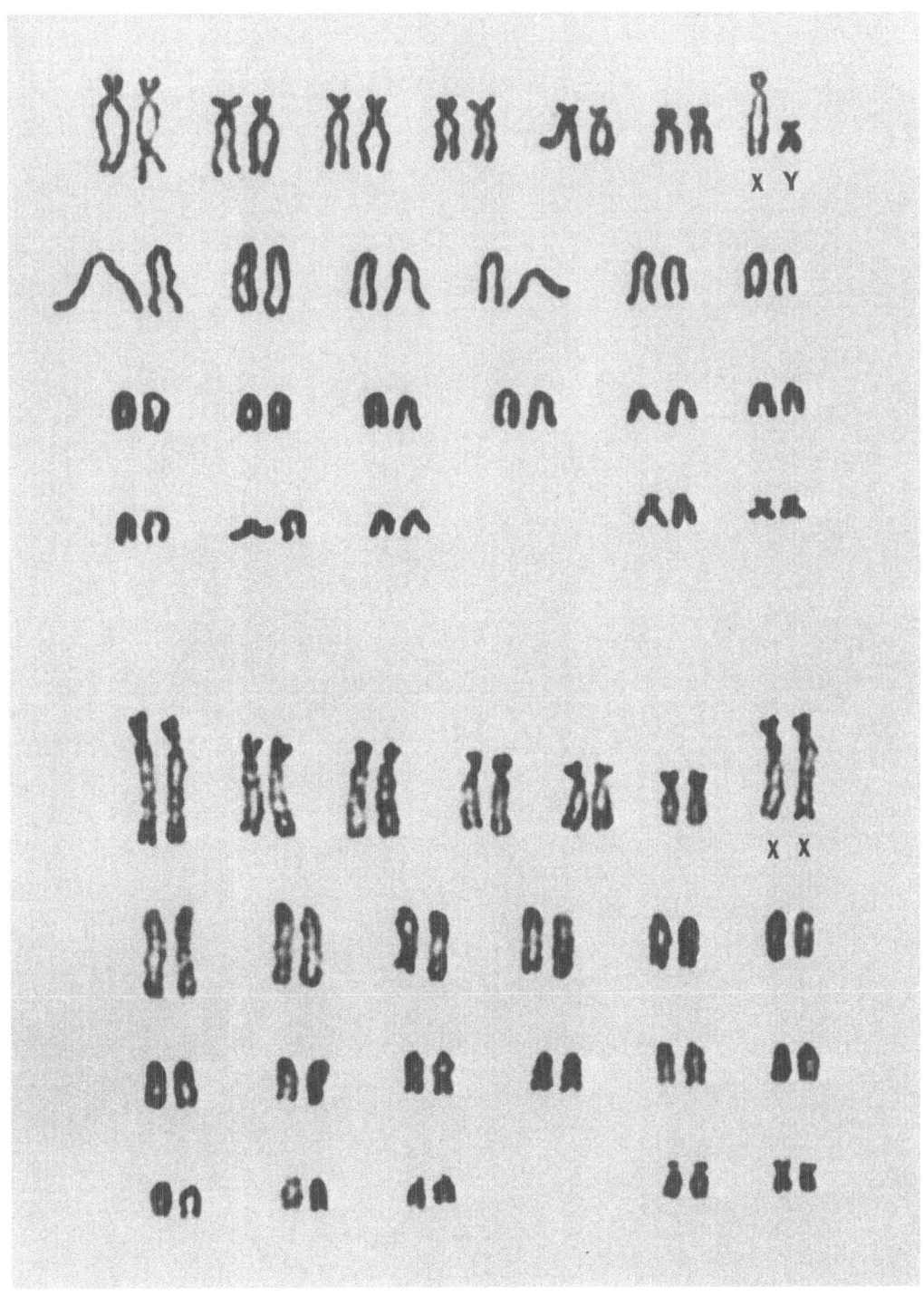

Order: RODENTIA

Family: CRICETIDAE

Peromyscus melanophrys (Plateau mouse)

$2n = 48$

Order: RODENTIA   Family: CRICETIDAE

<u>Peromyscus</u> <u>melanophrys</u> (Plateau mouse)

2n=48

AUTOSOMES: 10 Metacentrics and submetacentrics
             36 Acrocentrics

SEX CHROMOSOMES: X Submetacentric
                   Y Acrocentric

     According to the nomenclature system of Hsu and Arrighi, the biarmed
autosomes are, from left to right in the front row, A_1, A_2, A_3, B_1, B_2. From
a limited number of specimens analyzed, it is evident that polymorphism in
respect to chromosome features exists. The Y chromosome may vary in size,
and the number of biarmed autosomes may be higher than 10 in some individuals.

     The male specimen was collected by Dr. J. L. Patton in 1966 at
Zecatecas, Mexico, and the female was the second generation descendent of
animals originally caught by Dr. Rollin H. Baker in 1967 at 5 miles South
of Rio Grande, Murchison Ranch, Mexico. Lung cultures were initiated for
cytological preparations.

REFERENCES:

1) Hsu, T.C. and Arrighi, F.E.: Chromosomes of <u>Peromyscus</u> (Rodentia,
Cricetidae). I. Evolutionary trends in 20 species. Cytogenetics <u>7</u>:417,
1968.

<u>Peromyscus</u> <u>melanophrys</u> (Plateau mouse)

2n=48

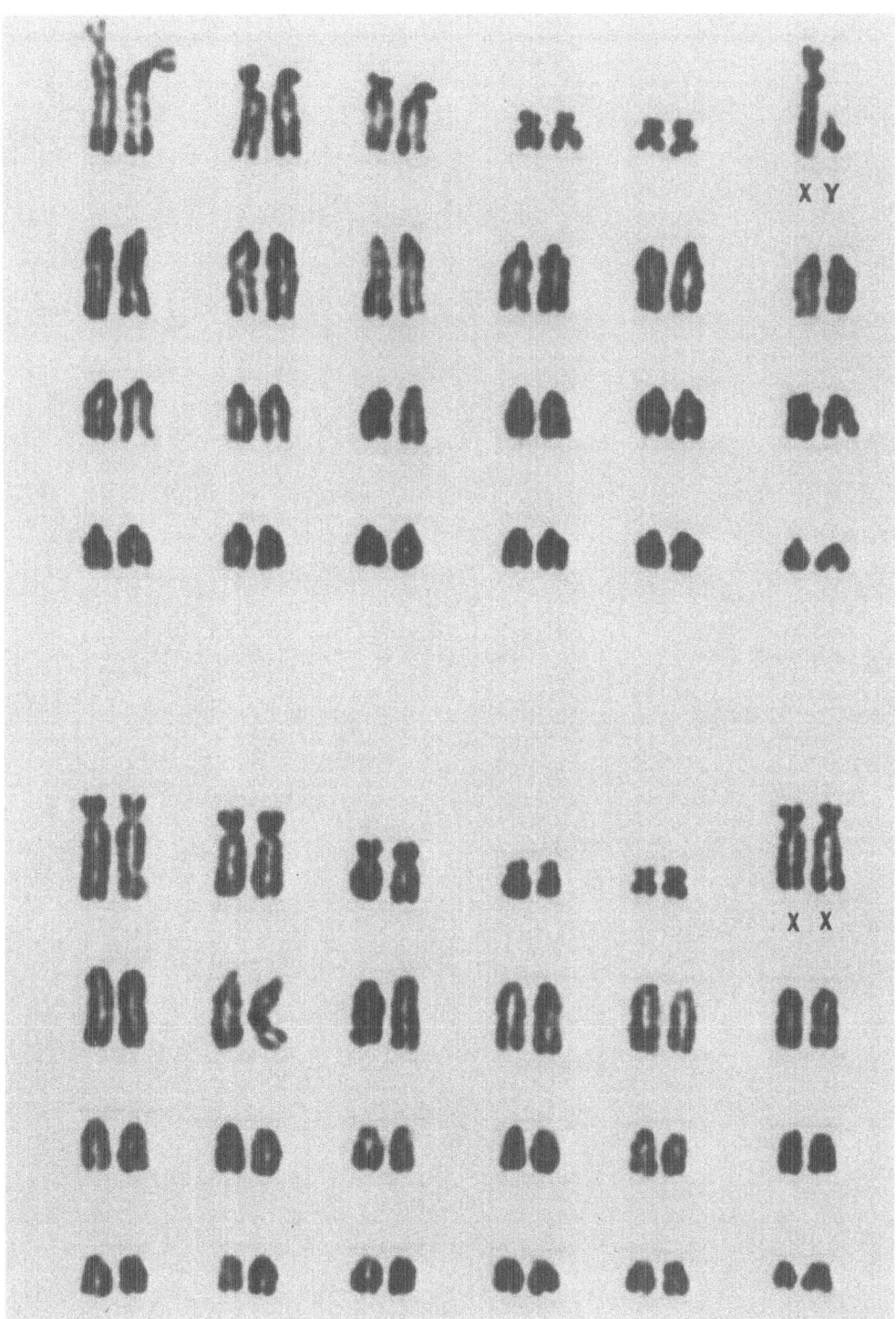

Order: RODENTIA

Family: CRICETIDAE

Arvicola terrestris (Water vole)

2n = 36

Order: RODENTIA   Family: CRICETIDAE

<u>Arvicola</u> <u>terrestris</u> (Water vole)

2n=36

AUTOSOMES: 26 Metacentrics and submetacentrics
              8 Acrocentrics

SEX CHROMOSOMES: X Submetacentric
                  Y Acrocentric

     These karyotypes were kindly donated by Dr. K. Fredga, Lund, Sweden,
and came from tissue cultures of various organs of five male and two female
specimens from different parts of Sweden. All were similar. Four
chromosomes may display secondary constrictions and satellites, and
satellite association occurs in acrocentrics. A trisomic animal has been
described by Fredga. The karyotype arrangement follows his description.

REFERENCES:

1) Matthey, R.: Analyse cytotaxonomique de huit espèces de Muridés.
<u>Murinae</u>, <u>Cricetinae</u>, <u>Microtinae</u> paléarctiques et nord-américains.
Arch. Klaus-Stift. Vererb.-Forsch. <u>32</u>:385, 1957.

2) Fredga, K.: Idiogram and trisomy of the water vole (<u>Arvicola</u>
<u>terrestris</u> L.), a favorable animal for cytogenetic research. Chromosoma <u>25</u>:
75, 1968.

Order: RODENTIA

Family: CRICETIDAE

Arvicola terrestris (Water vole)

2n=36

Order: RODENTIA

Family: CRICETIDAE

Lagurus curtatus (Sagebrush vole)
$2n = 54$

Lagurus curtatus (Sagebrush vole)

2n=54

AUTOSOMES: 2 Submetacentrics
            50 Acrocentrics

SEX CHROMOSOMES: X Submetacentric
                  Y Acrocentric

Identification of the X chromosomes is unequivocal, but identification of the Y is subjective. Among the acrocentric autosomes, only one pair, the longest, can be separated from the rest.

The specimens were kindly donated by Dr. Murray L. Johnson. The male was collected in Jefferson County, Oregon, USA, and the female, from East Mansfield, Douglas County, Washington, USA. Lung cultures were initiated for cytological preparations.

Order: RODENTIA

Family: CRICETIDAE

Lagurus curtatus (Sagebrush vole)

2n=54

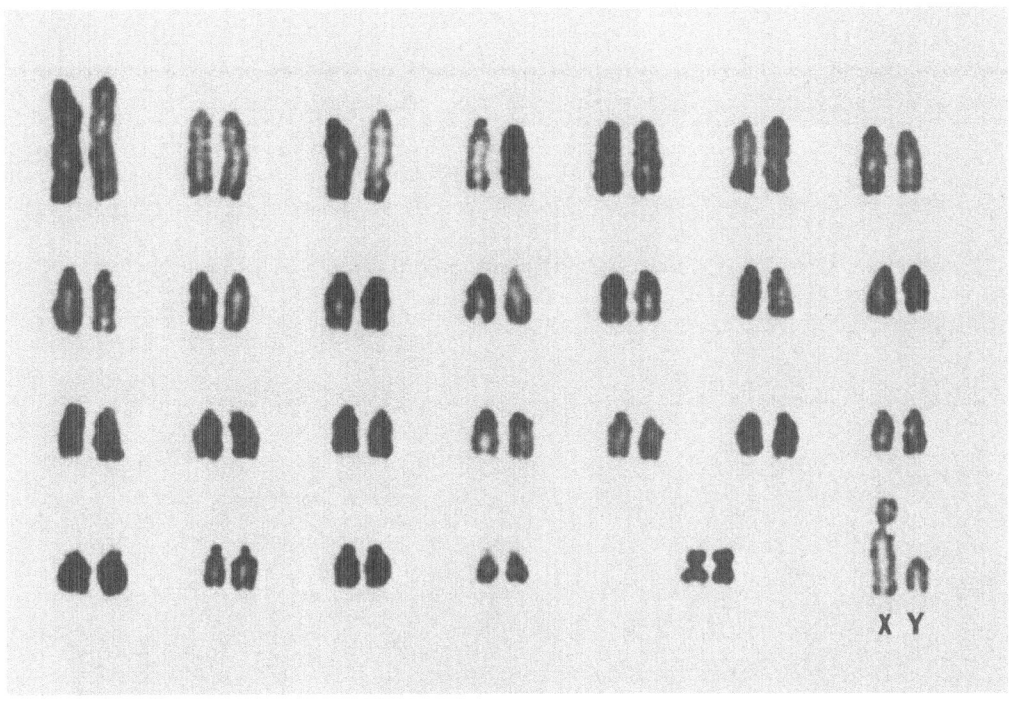

X Y

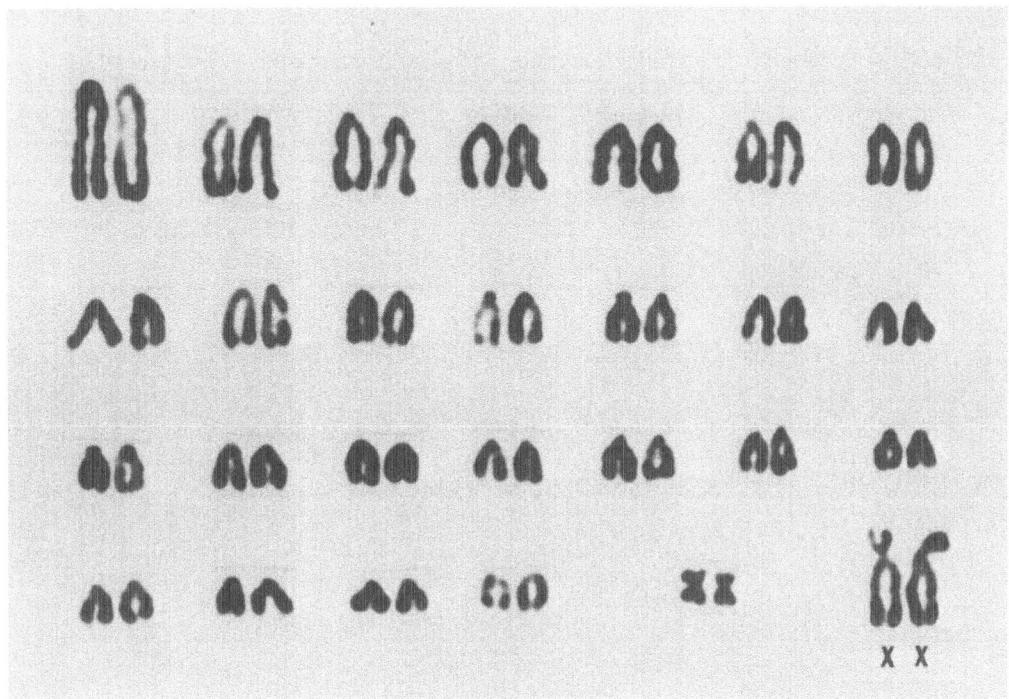

X X

Order: RODENTIA

Family: CRICETIDAE

Lagurus lagurus (Steppe lemming)

$2n = 54$

<u>Lagurus lagurus</u> (Steppe lemming)

2n=54

AUTOSOMES: 8 Metacentrics and submetacentrics
             44 Acrocentrics

SEX CHROMOSOMES: X Submetacentric
                  Y Acrocentric

  All the four pairs of biarmed chromosomes are morphologically
distinguishable. The identification of the X chromosome is unequivocal;
but the identification of the Y is not. The Y is apparently one of the
smallest acrocentrics.

  The karyotypes presented here are gifts of Dr. K. Kernahle, Department
of Biology, Martin-Luther-Universität, DDR402 Halle/S., Universitätsplatz 7.,
Germany. The animals were descendents of four pairs imported in 1963 from
Moscow, USSR.

REFERENCES:

1) Matthey, R.: Analyse cytotaxonomique de huit espèces de Muridés. <u>Murinae</u>
<u>Cricetinae</u>, <u>Microtinae</u> paléarctiques et nord-américains. Arch. Julius
Klaus-Stiftung <u>32</u>:26, 1957.

2) Matthey, R.: Les chromosomes des Mammifères euthériens. Arch.
Julius-Klaus-Stiftung <u>33</u>:253, 1958.

3) Pogosianz, H.E. und A.F. Zaharov: On the karyotype of the steppe lemming
(<u>Lagurus lagurus</u> Pallas). Z. Versuchstierkd. <u>1</u>:93, 1962.

4) Zernahle, K.: Zur Zytogenetik des Steppenlemmings (<u>Lagurus lagurus</u>
Pallas). Der Kerngeschlechtdimorphismus und die Chromosomen in der Mitose.
In Vorbereitung, 1969.

Order: RODENTIA

Lagurus lagurus (Steppe lemming)

2n=54

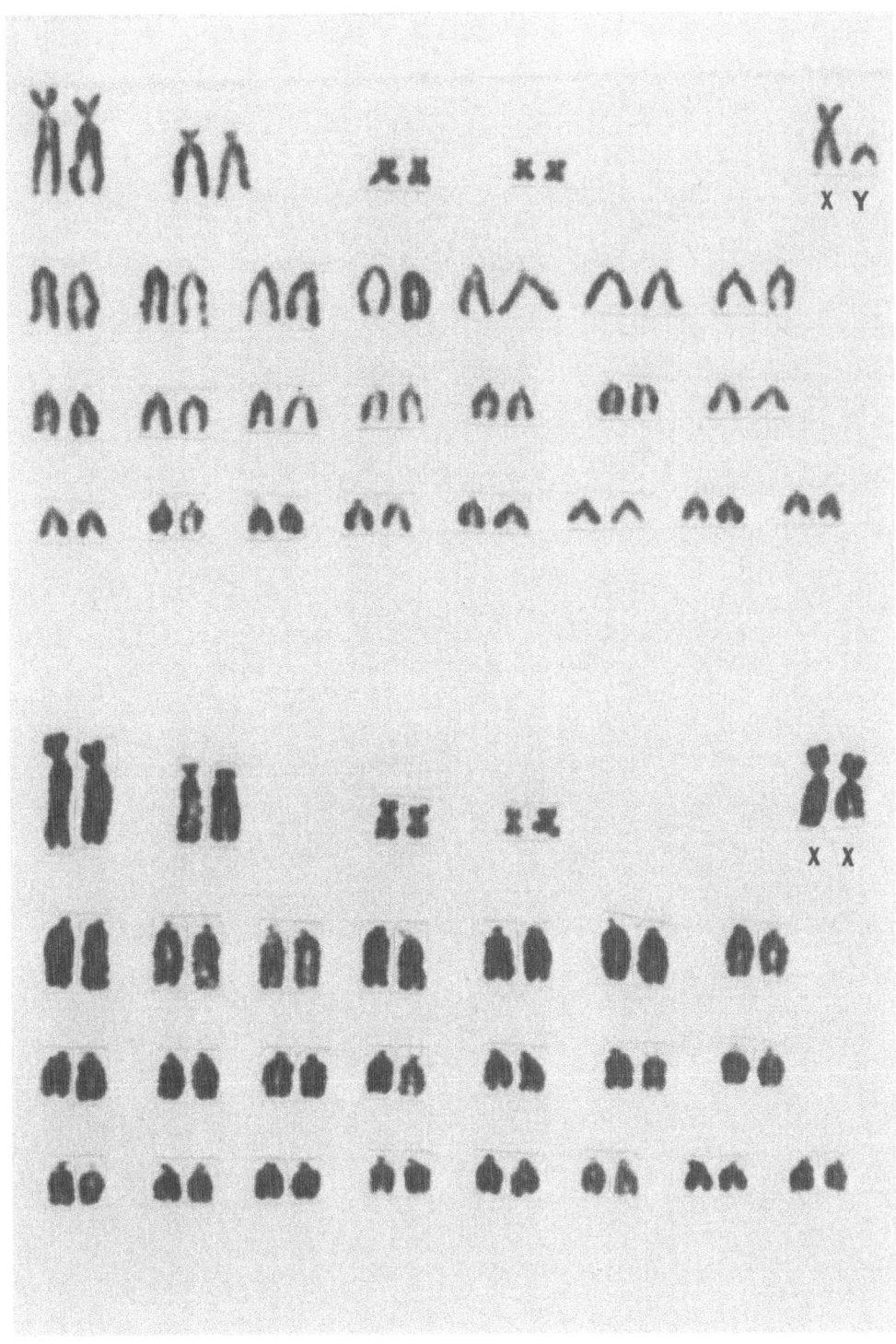

Order: RODENTIA

Family: CRICETIDAE

Microtus townsendii (Townsend's vole)
$2n = 50$

Microtus townsendii (Townsend's vole)

2n=50

AUTOSOMES: 48 Acrocentrics

SEX CHROMOSOMES: X Acrocentric
                   Y Acrocentric

The entire complement forms a smooth gradation of sizes without a significant break, so that pairing chromosomes is extremely subjective. The X is apparently one of the larger elements, and the Y, one of the smaller elements.

The specimens (2♀♀, 2♂♂) were kindly donated by Dr. J. Mary Taylor, University of British Columbia, Vancouver, B.C., Canada. They were trapped in the campus of the University of British Columbia. Lung cultures were initiated for cytological preparations. Two additional specimens, 1♀, 1♂, from Pierce County, Washington, USA, supplied by Dr. Murray L. Johnson, gave similar results.

Microtus townsendii (Townsend's vole)

2n=50

Order: RODENTIA

Family: CRICETIDAE

Ondatra zibethica (Muskrat)

$2n = 54$

Ondatra zibethica (Muskrat)

2n=54

AUTOSOMES: 50 Acrocentrics
             2 Small submetacentrics

SEX CHROMOSOMES: X Acrocentric
                   Y Acrocentric

Among the acrocentric autosomes at least one pair is satellited. This is clearly demonstrated in the male karyotype here. The X chromosomes are the longest acrocentrics. This conclusion is ascertained not only from morphology, but also from tritiated thymidine autoradiography.

The karyotypes presented here are gifts of Professor Alfred Gropp. The animals were trapped near the mouth of the Sieg River in the Rhine, approximately 10 miles from Bonn, Germany.

REFERENCES:

1) Matthey, R.: Nouvelles recherches sur les chromosomes des Muridae. Caryologia 6:1, 1954.

2) Moore, W., Jr., Elder, R.L. and Gillespie, L.J.: The chromosomes of the muskrat. J. Hered. 57:104, 1966.

3) Gropp, A. and Geisler, M.: Chromosomes of the muskrat (Ondatra zibethica L., Muridae, Rodentia). Mammalian Chromosomes Newsletter 8:286, 1967.

4) Sutton, D.A.: Chromosomes of the muskrat (Ondatra zibethica L., Microtinae, Rodentia). Mammalian Chromosomes Newsletter 9:244, 1969.

5) Sutton, D.A.: Chromosomes of a male muskrat (Ondatra zibethica L., Microtinae, Rodentia). Mammalian Chromosomes Newsletter 11:24, 1970.

6) Dewald, G.W. and Jalal, S.M.: Karyotypes of two subspecies of North American muskrat. Mammalian Chromosomes Newsletter 11:37, 1970.

Ondatra zibethica (Muskrat)

2n=54

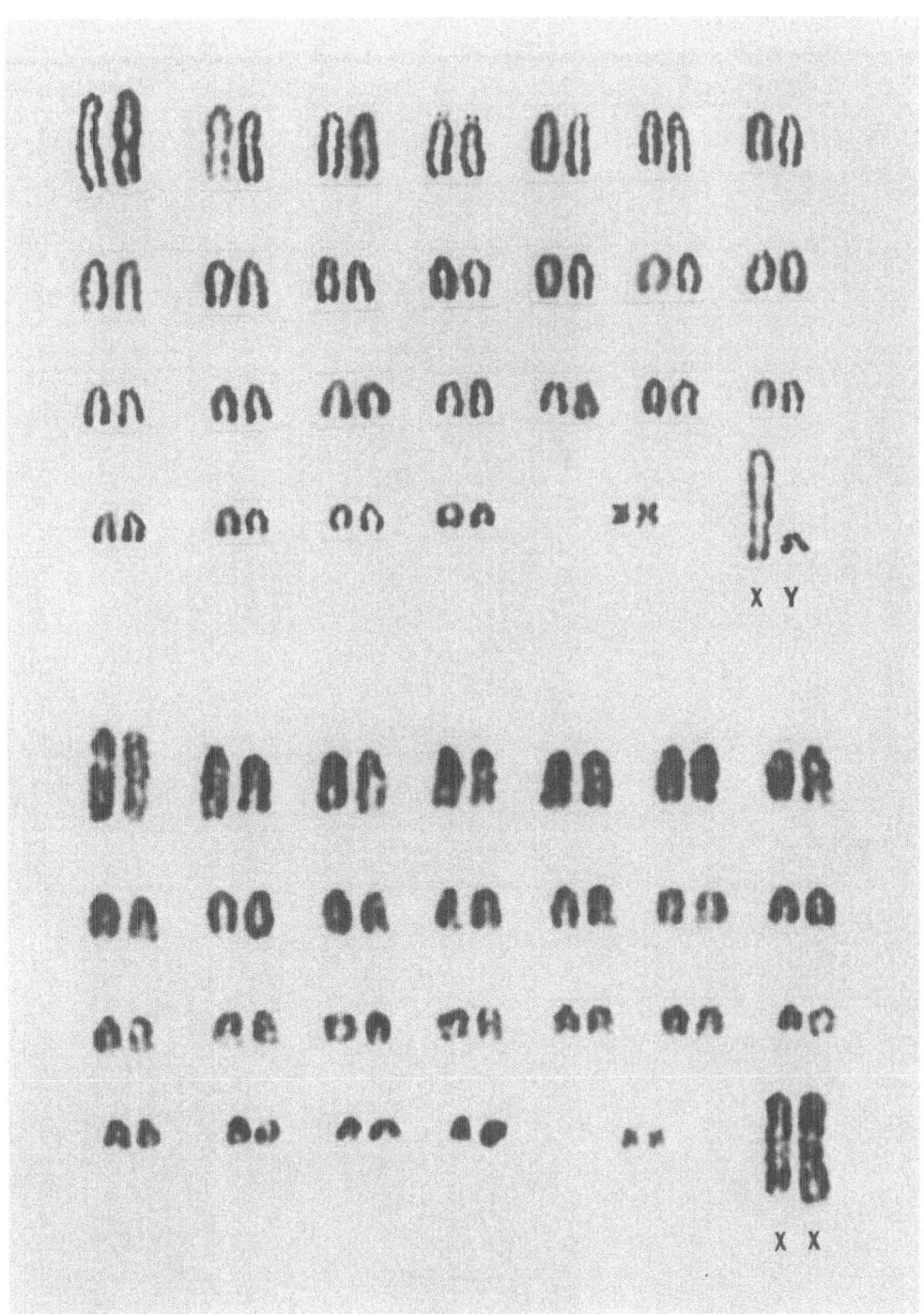

Order: RODENTIA

Family: CRICETIDAE

Meriones hurrianae (Indian desert gerbil)

$2n = 40$

Order: RODENTIA   Family: CRICETIDAE

<u>Meriones</u> <u>hurrianae</u> (Indian desert gerbil)

2n=40

AUTOSOMES: 36 Metacentrics and submetacentrics
              2 Acrocentrics

SEX CHROMOSOMES: X Submetacentric
                  Y Submetacentric

    Many autosome pairs are morphologically similar and they form a more
or less continuous size gradation. The X is one of the longest chromosomes
in the complement.

    The karyotypes presented here are gifts of Dr. Charles F. Nadler. The
animals were caught in Gizri (2 miles E. Karachi), Pakistan. Bone marrows
were used for cytological preparations.

REFERENCES:

1) Nadler, C.F. and Lay, D.M.: Chromosomes of some species of <u>Meriones</u>
(Mammalia: Rodentia). Z. f. Säugetierkunde <u>32</u>:285, 1967.

Meriones hurrianae (Indian desert gerbil)

2n=40

Order: RODENTIA

Family: CRICETIDAE

Meriones libycus (Libyan jird)
$2n = 44$

<u>Meriones libycus</u> (Libyan jird)

2n=44

AUTOSOMES: 30 Metacentrics and submetacentrics
             12 Acrocentrics

SEX CHROMOSOMES: X Acrocentric
                   Y Submetacentric

    The long acrocentric X chromosomes are characteristic. It can be used
as an excellent marker in cytological studies in hybrids between <u>M</u>. <u>shawi</u> and
<u>M</u>. <u>libycus</u>. The two species also differ in several autosomal characteristics,
though the diploid numbers are the same. For <u>M</u>. <u>shawi</u>, cf. Folio 227.

    The karyotypes presented here are gifts of Dr. Charles F. Nadler. The
specimens were from laboratory stock originally obtained from Lichtenstein-
Rachtagan, Iran.

REFERENCES:

1) Matthey, R.: Les Chromosomes des Muridae. Rev. Suisse de Zool. <u>60</u>:
225, 1953.

2) Matthey, R.: Cytologie et taxonomie du genre <u>Meriones</u>, Illiger.
Säugetierk. Mitt. <u>5</u>:145, 1957.

3) Nadler, C.F. and Lay, D.M.: Chromosomes of some species of <u>Meriones</u>
(Mammalia: Rodentia). Z. f. Säugetiekunde <u>32</u>:285, 1967.

4) Lay, D.M. and Nadler, C.F.: Hybridization in the rodent genus <u>Meriones</u>.
I. Breeding and cytological analyses of <u>Meriones</u> <u>shawi</u> (♀) X <u>Meriones</u> <u>libycus</u>
(♂) hybrids. Cytogenetics <u>8</u>:35, 1969.

5) Nadler, C.F.: Chromosomal Evolution in Rodents. In "Comparative
Mammalian Cytogenetics" (Benirschke, K., ed.), pp. 277-309, Springer-Verlag,
New York, 1969.

Meriones libycus (Libyan jird)

2n=44

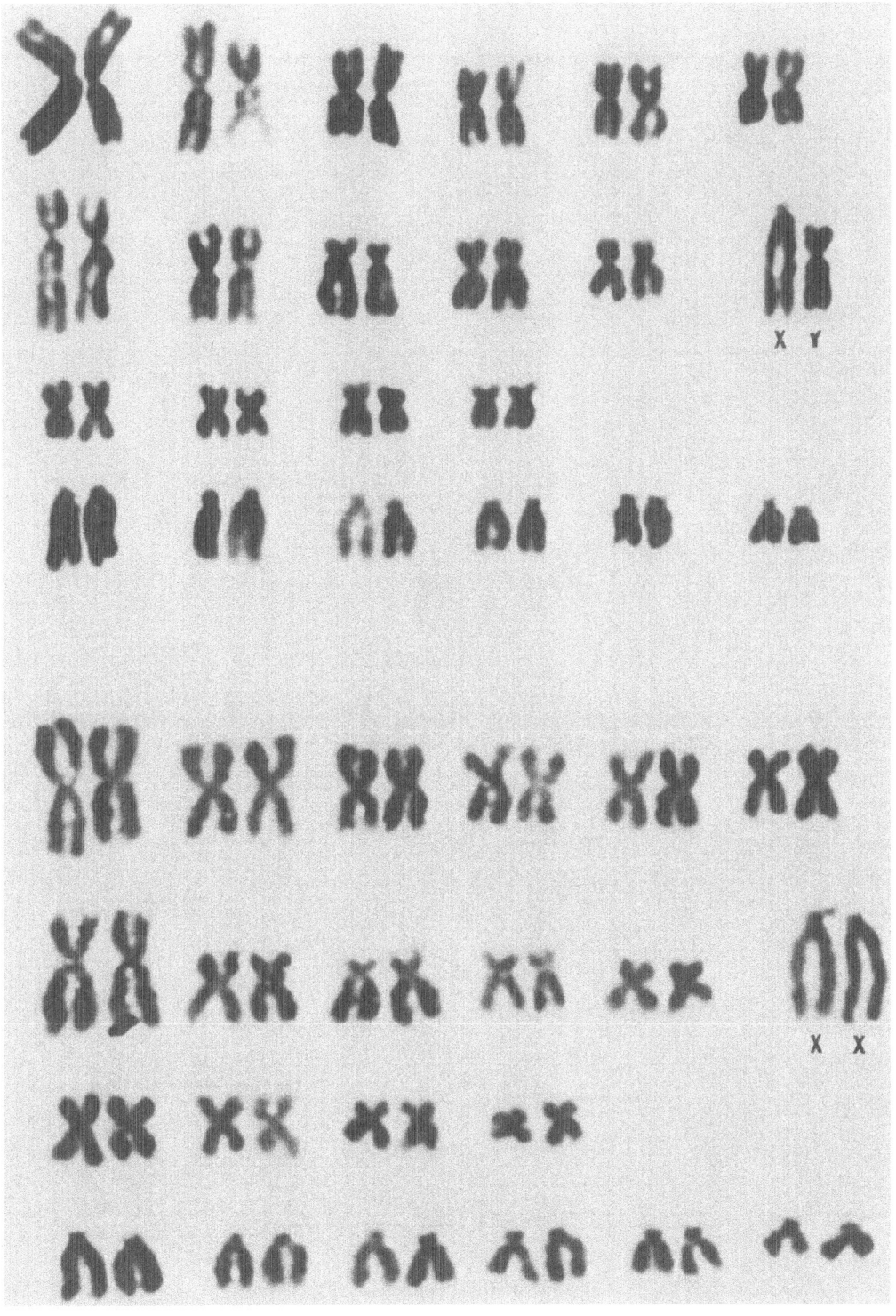

Order: RODENTIA

Family: CRICETIDAE

Meriones shawi (Shaw's jird)
$2n = 44$

Meriones shawi (Shaw's jird)

2n=44

AUTOSOMES: 30 Metacentrics and submetacentrics
             12 Acrocentrics

SEX CHROMOSOMES: X Submetacentric
                   Y Submetacentric

The large submetacentric X chromosome is an excellent distinguishing marker when the karyotypes of M. shawi are compared with those of M. libycus (Folio 226). Several differences also exist among the autosomes between the two species. One of the acrocentric pairs has distinct short arm. This species can be hybridized with M. libycus.

The karyotypes presented here are gifts of Charles F. Nadler. The specimens were from laboratory stock originally obtained from Duvernoy-Bahig, Burgel Arab, Egypt.

REFERENCES:

1) Matthey, R.: Les Chromosomes des Muridae. Rev. Suisse de Zool. 60: 225, 1953.

2) Matthey, R.: Cytologie et taxonomie du genre Meriones, Illiger. Säugetierk. Mitt. 5:145, 1957.

3) Nadler, C.F. and Lay, D.M.: Chromosomes of some species of Meriones (Mammalia: Rodentia). Z. f. Säugetierkunde 32:285, 1967.

4) Lay, D.M. and Nadler, C.F.: Hybridization in the rodent genus Meriones. I. Breeding and cytological analyses of Meriones shawi (♀) X Meriones libycus (♂) hybrids. Cytogenetics 8:35, 1969.

5) Nadler, C.F.: Chromosomal Evolution in Rodents. In "Comparative Mammalian Cytogenetics" (Benirschke, K., ed.), pp. 277-309, Springer-Verlag, New York, 1969.

Order: RODENTIA

Family: CRICETIDAE

Meriones shawi (Shaw's jird)

2n=44

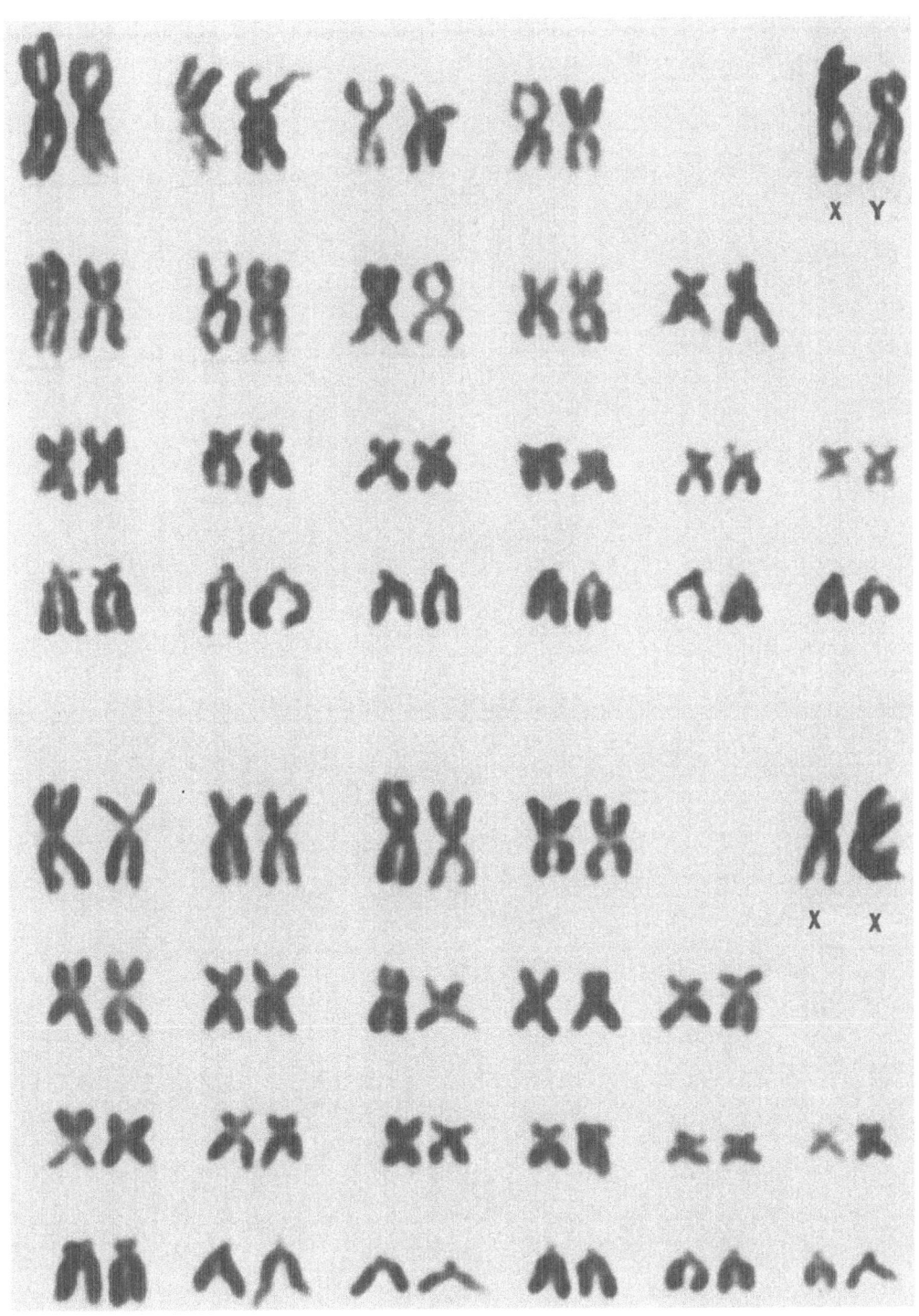

Volume 5, Folio 227, 1971

Order: RODENTIA

Family: MURIDAE

Mus cervicolor

$2n = 40$

Mus cervicolor

2n=40

AUTOSOMES: 38 Acrocentrics

SEX CHROMOSOMES: X Acrocentric
                  Y Acrocentric

The karyotype of this species is indistinguishable from that of
M. musculus (Folio 17). Pairing of the chromosomes is arbitrary.

The specimens were laboratory descendents of specimens originally
caught in Thailand and were donated by Dr. Frank H. Ruddle, Yale University,
New Haven, Connecticut, USA. The subspecies is M. c. caroli.

Mus cervicolor

2n=40

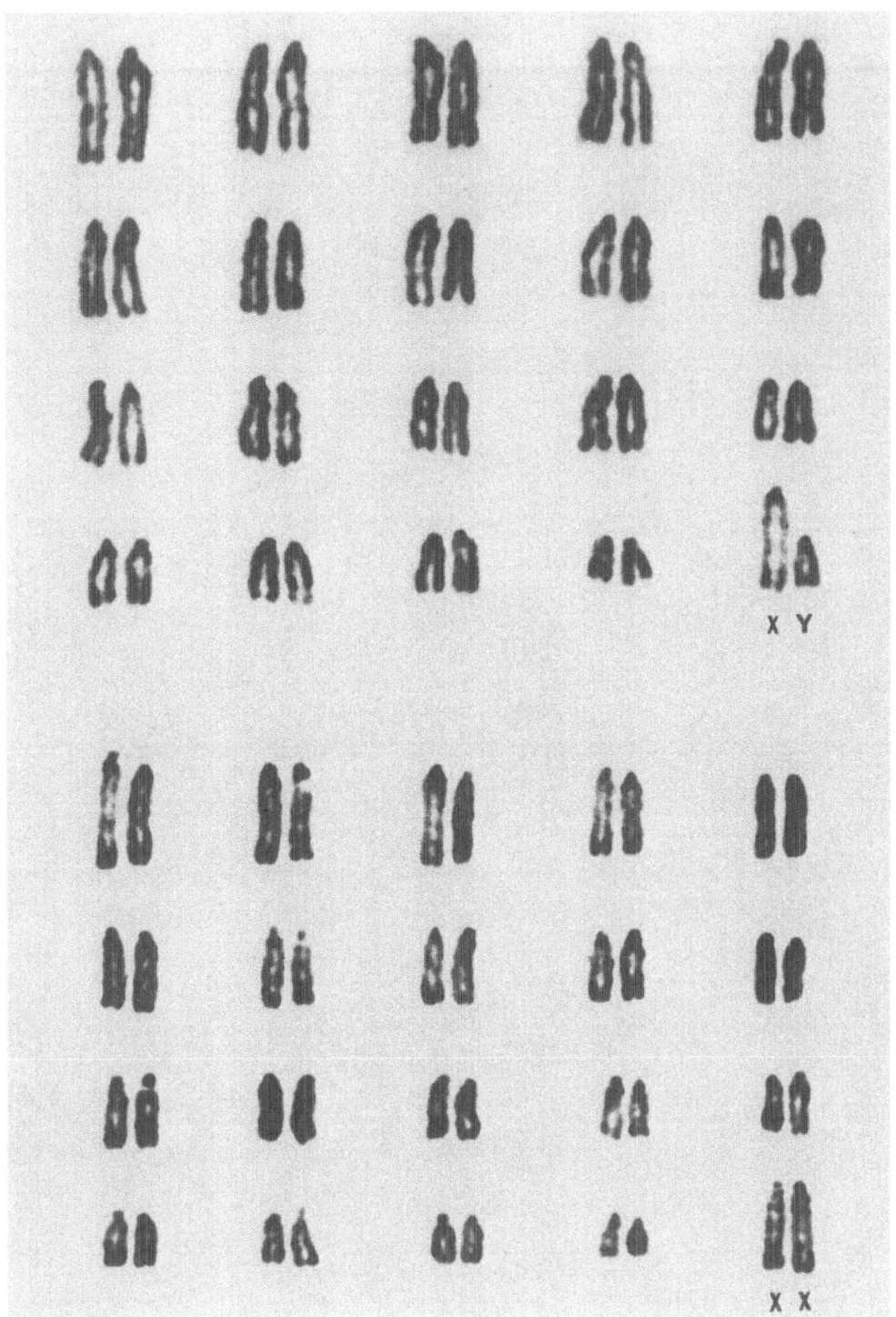

Order: RODENTIA

Family: MURIDAE

Rattus rattus (Black rat)

$2n = 42$

Order: RODENTIA   Family: MURIDAE

<u>Rattus</u> <u>rattus</u> (Black rat)

2n=42

AUTOSOMES: 18 Metacentrics and submetacentrics
             22 Acrocentrics

SEX CHROMOSOMES: X Acrocentric
                   Y Acrocentric

All the metacentrics and submetacentrics are either medium-size or small elements. Some of the acrocentrics, especially two pairs, may be long. The X chromosome is among the long acrocentrics, but the identification of the Y offers no problem.

The specimens were collected by Mr. John F. Duncan and his collaborators in South Vietnam. The male was trapped in Calu, and the female, in Quang Tri Combat Base, bearing specimen numbers RHL-58 and DVP 468, respectively. The specimens are stored in the Smithsonian Institute, Washington, D.C., USA. Lung cultures were used for cytological preparations.

Polymorphism exists in many wild populations in the Orient. Yosida and his associates found that the longest autosomal pair and the 9th chromosome may be acrocentric, submetacentric, or hemizygous of these. Also, wild specimens may possess an additional small acrocentric autosome in the complement to give 43 as the diploid number. In the Vietnam samples, such a chromosome was also found in a few individuals.

REFERENCES:

1) Yosida, T.H., Nakamura, A. and Fukaya, T.: Chromosomal polymorphism in <u>Rattus</u> <u>rattus</u> (L.) collected in Kusudomari and Misima. Chromosoma <u>16</u>:70, 1965

2) Yosida, T.H. and Associates: Many reports in Ann. Rep. Nat. Inst. Genet., Japan, 1965-1969.

l) Yosida, T.H., Tsuchiya, K., Imai, H.T. and Morwaki, K.: New chromosome ypes of the black rat, <u>Rattus</u> <u>rattus</u>, collected in Oceania and hybrids etween Japanese and Australian rats. Jap. J. Genet. <u>44</u>:89, 1969.

) Yosida, T.H., Tsuchiya, K., Imai, H., Moriwaki, K. and Udagawa, T.: hromosome numbers of rodent species in South East Asia and Oceania. ammalian Chromosomes Newsletter <u>10</u>:217, 1969.

Yong, H.S.: Karyotypes of rats from Hong Kong and Thailand (Muridae, nus <u>Rattus</u> Fischer). Cytologia <u>34</u>:394, 1969.

Yong, H.S.: Karyotypes of Malayan rats (Rodentia-Muridae, genus <u>Rattus</u> scher). Chromosoma <u>27</u>:245, 1969.

Capanna, E. and Ciritelli, M.V.: An endemic population of <u>Rattus</u> <u>rattus</u>) with a 38-chromosome complement. Mammalian Chromosomes Newsletter <u>10</u>: , 1969.

Rattus rattus (Black rat)

2n=42

Order: CARNIVORA

Family: CANIDAE

Nyctereutes procyonoides (viverrinus) (Raccoon dog)

2n = 42

Nyctereutes procyonoides (viverrinus) (Raccoon dog)

2n=42

AUTOSOMES: 26 Metacentrics and submetacentrics
             14 Acrocentrics

SEX CHROMOSOMES: X Acrocentric
                  Y Acrocentric with satellites

     Skin biopsies of a male and a female were kindly provided by
Dr. C. Gray, National Zoological Park, Washington, D.C., USA. The male
was mosaic for 42/43 chromosomes, apparently XY/XXY, although testicular
biopsy disclosed normal seminiferous tubules. The X chromosome was
identified by autoradiography.

REFERENCES:

1) Minouchi, O.: On the spermatogenesis of the raccoon dog (Nyctereutes
viverrinus), with special reference to the sex chromosomes. Cytologia 1:
88, 1929.

2) Wurster, D.H.: Cytogenetic and phylogenetic studies in Carnivora.
In Comparative Mammalian Cytogenetics (Benirschke, K., ed.), Springer-Verlag,
New York, 1969.

3) Todd, N.B. and Pressman, S.R.: The karyotype of the raccoon dog
(Nyctereutes sp.). Mammalian Chromosomes Newsletter 10:21, 1969.

Nyctereutes procyonoides (viverrinus) (Raccoon dog)

2n=42

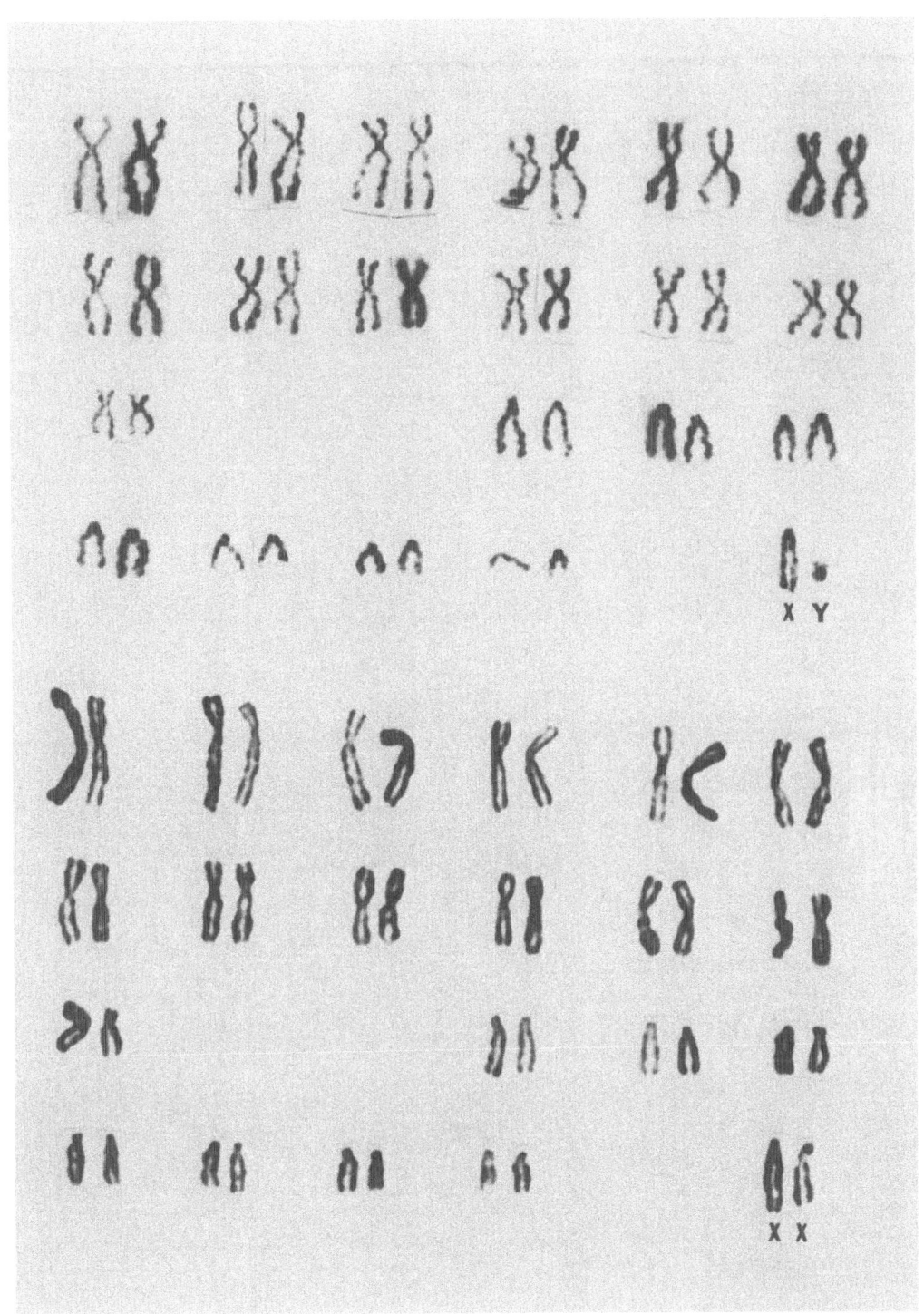

Order: CARNIVORA

Family: MUSTELIDAE

Mustela frenata (Long-tailed weasel)

$2n = 42$

Order: CARNIVORA   Family: MUSTELIDAE

<u>Mustela</u> <u>frenata</u> (Long-tailed weasel)

2n=42

AUTOSOMES: 20 Metacentrics and submetacentrics
              20 Acrocentrics

SEX CHROMOSOMES: X Submetacentric
                  Y Submetacentric

    Metaphases for these karyotypes were kindly donated by Dr. P. K. Basrur, Guelph, Ontario, Canada, and came from blood and fibroblast cultures of three male and two female animals trapped in Southern Ontario. The animals were identified by features such as skull measurements, and Basrur gives the measurements of arm ratios. She considers there to be only 18 acrocentrics, the second pair to last autosomes shown here were considered as biarmed elements. A secondary constriction is regularly found near the ends of the long arms of the last element in the second row and less commonly in the same position of the elements shown here in position two.

REFERENCES:

1) Basrur, P.K.: The karyotype of the long-tailed weasel <u>Mustela</u> <u>frenata</u> <u>noveboracensis</u> Emmons. Canad. J. Genet. Cytol. <u>10</u>:390, 1968.

<u>Mustela</u> <u>frenata</u> (Long-tailed weasel)

2n=42

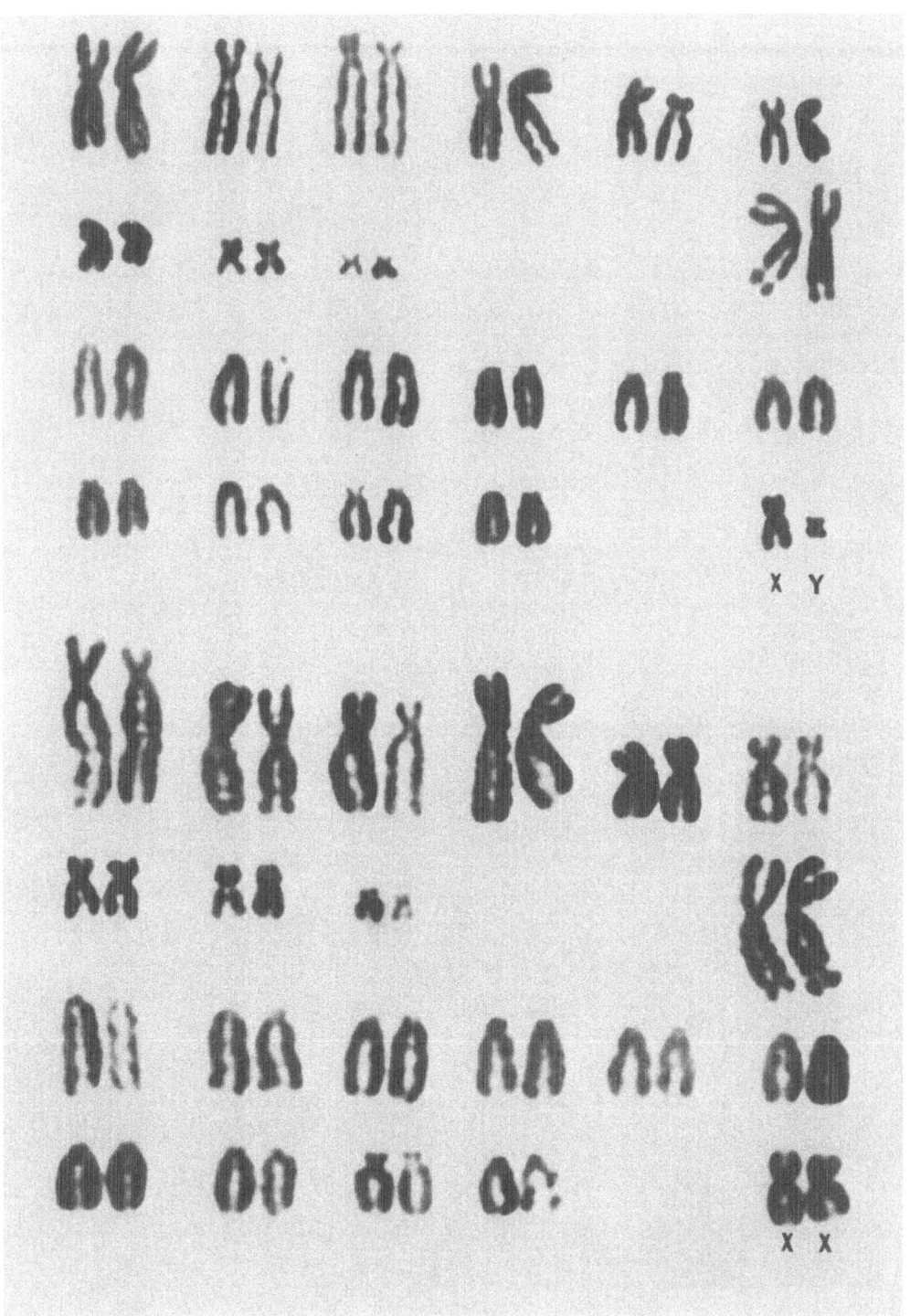

Order: CARNIVORA

Family: VIVERRIDAE

Bdeogale sp. (Black-footed mongoose)

2n = 36

Bdeogale sp. (Black-footed mongoose)

2n=36

AUTOSOMES: 28 Metacentrics and submetacentrics
              6 Acrocentrics

SEX CHROMOSOMES: X Submetacentric
                   Y Submetacentric

     The size of the short arms of No. 5 appears to be variable. No
chromosome pair showed conspicuous secondary constrictions found as good
marker chromosomes in many carnivore karyotypes.

     Skin biopsies were kindly made available from one animal of each sex
by Dr. C. Gray, National Zoological Park, Washington, D.C., USA.

REFERENCES:

1) Wurster, D.H. and Benirschke, K.: Chromosome numbers in thirty species
of carnivores. Mammalian Chromosomes Newsletter 8:195, 1967.

2) Wurster, D.H. and Benirschke, K.: Comparative cytogenetic studies in
the Order Carnivora. Chromosoma 24:336, 1968.

<u>Bdeogale</u> <u>sp</u>. (Black-footed mongoose)

2n=36

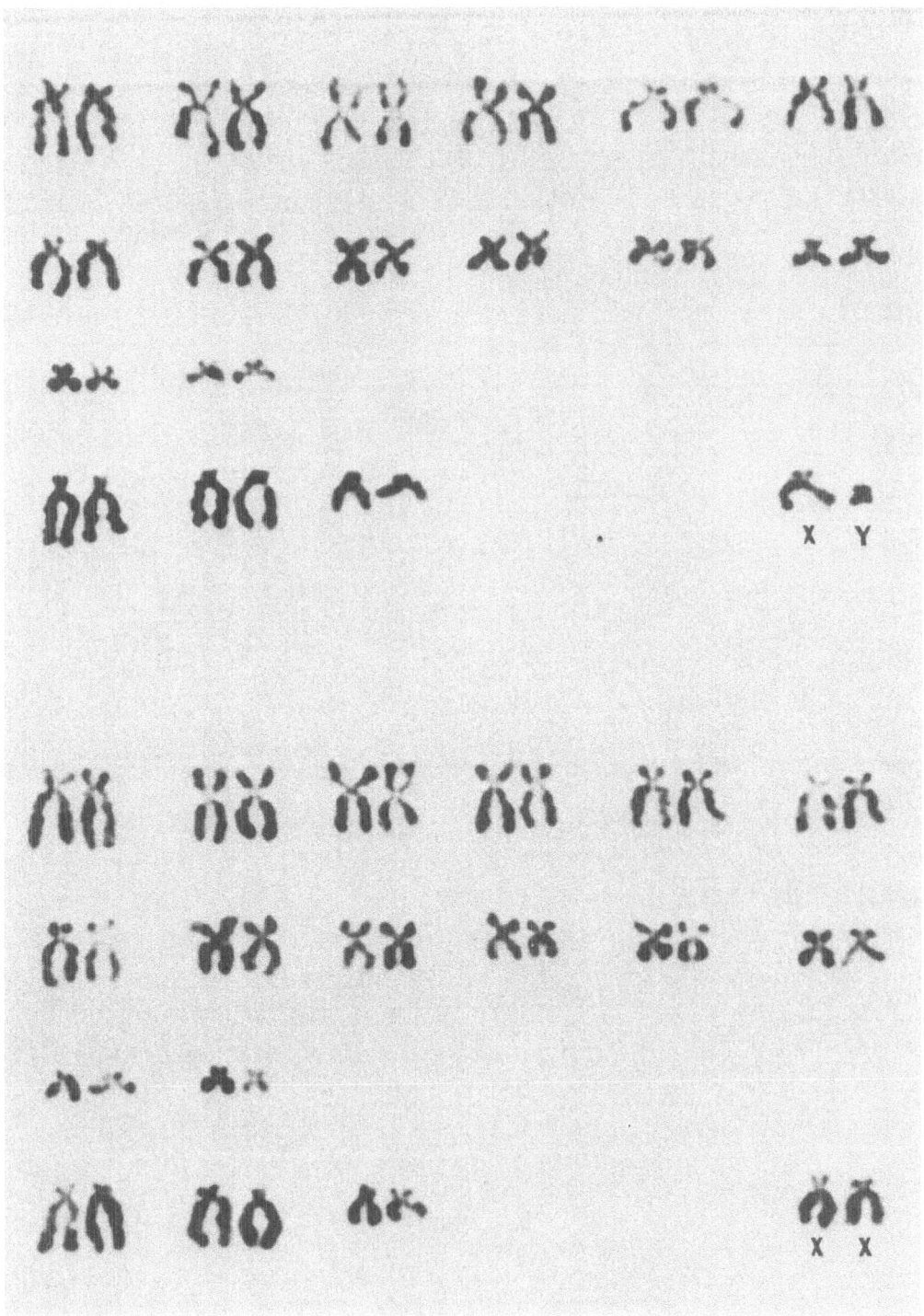

Order: CARNIVORA

Family: VIVERRIDAE

Paradoxurus hermaphroditus cochinensis (Palm civet)

$2n = 42$

Paradoxurus hermaphroditus cochinensis (Palm civet)

2n=42

AUTOSOMES: 22 Metacentrics and submetacentrics
             18 Acrocentrics

SEX CHROMOSOMES: X Submetacentric
                   Y Minute acrocentric

The marker chromosome pair with pronounced secondary constrictions and satellites is here placed as element No. 7.

Skin biopsies of one sex each were kindly provided by Mr. J. F. Duncan from animals captured in South Vietnam.

REFERENCES:

1) Ray-Chaudhuri, S.P., Ranjini, P.V. and Sharma, T.: Somatic chromosomes of the common palm civet, Paradoxurus hermaphroditus. Experientia 22:740, 1966.

Order: CARNIVORA

Family: VIVERRIDAE

Paradoxurus hermaphroditus cochinensis (Palm civet)

2n=42

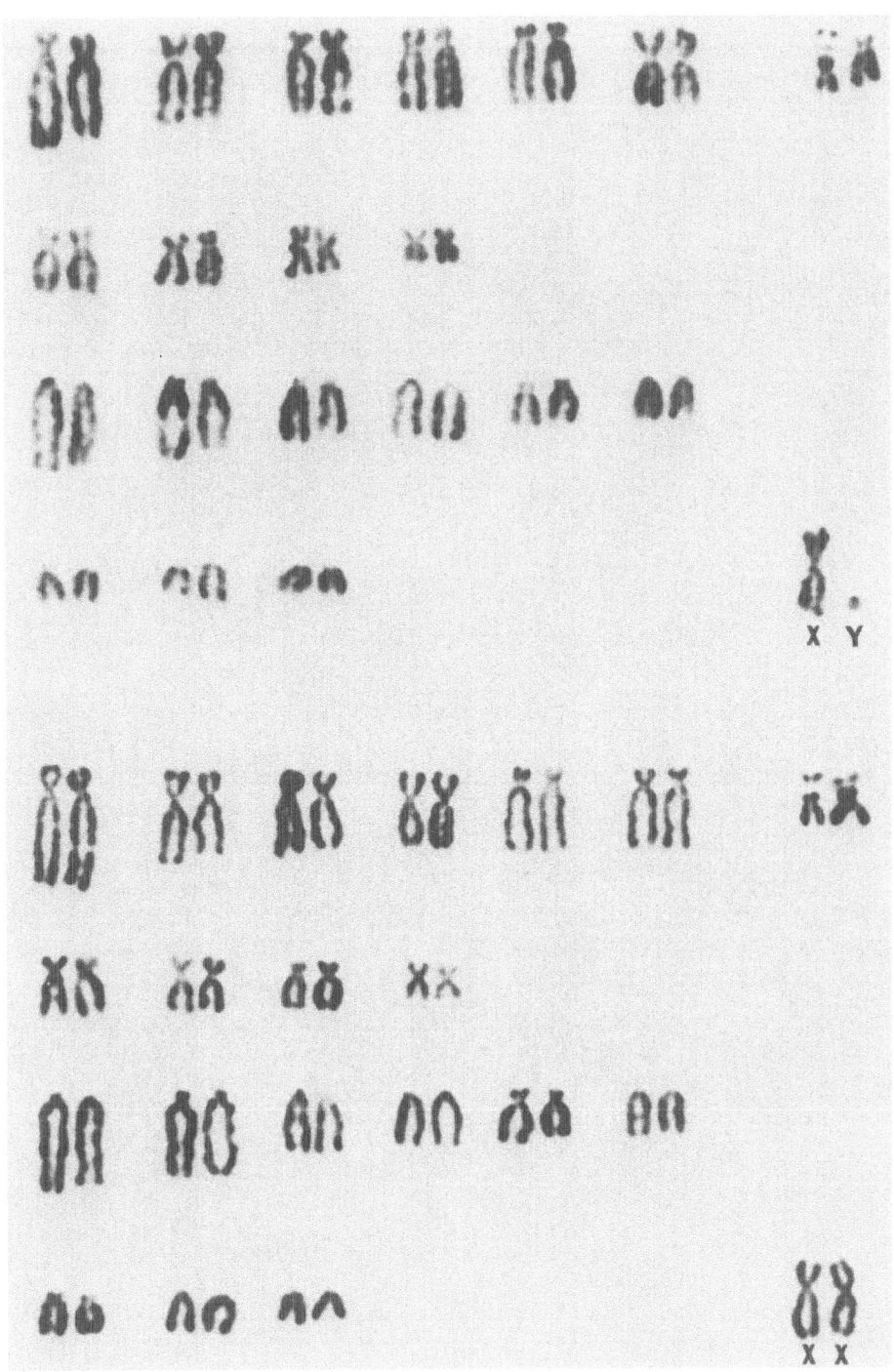

Order: CARNIVORA

Family: FELIDAE

Acinonyx jubatus (Cheetah)

$2n = 38$

<u>Acinonyx jubatus</u> (Cheetah)

2n=38

AUTOSOMES: 34 Metacentrics and submetacentrics
             2 Acrocentrics

SEX CHROMOSOMES: X Submetacentric
                   Y Acrocentric

     According to the San Juan nomenclature system for Felidae (see Folio 31) the autosomes are composed of three pairs in group A, four pairs in group B, two pairs in group C, four pairs in group D, four pairs in group E, and one pair in group F. The satellited pair is E_1.

     The male karyotype was donated by Dr. W. L. Hard, Omaha, Nebraska, USA, and was prepared from a lymphocyte culture of an animal imported from Somaliland and located at H. Doorley Zoo, Omaha. The female karyotype came from skin fibroblast cultures taken by Dr. C. Gray, National Zoological Park, Washington, D.C., USA.

REFERENCES:

1) Hsu, T.C., Rearden, H.H. and Luquette, G.F.: Karyological studies of nine species of Felidae. Amer. Naturalist <u>97</u>:225, 1963.

2) Hard, W.L.: The karyotype of a male cheetah, <u>Acinonyx jabatus jabatus</u>. Mammalian Chromosomes Newsletter <u>9</u>:16, 1968.

Acinonyx jubatus (Cheetah)

2n=38

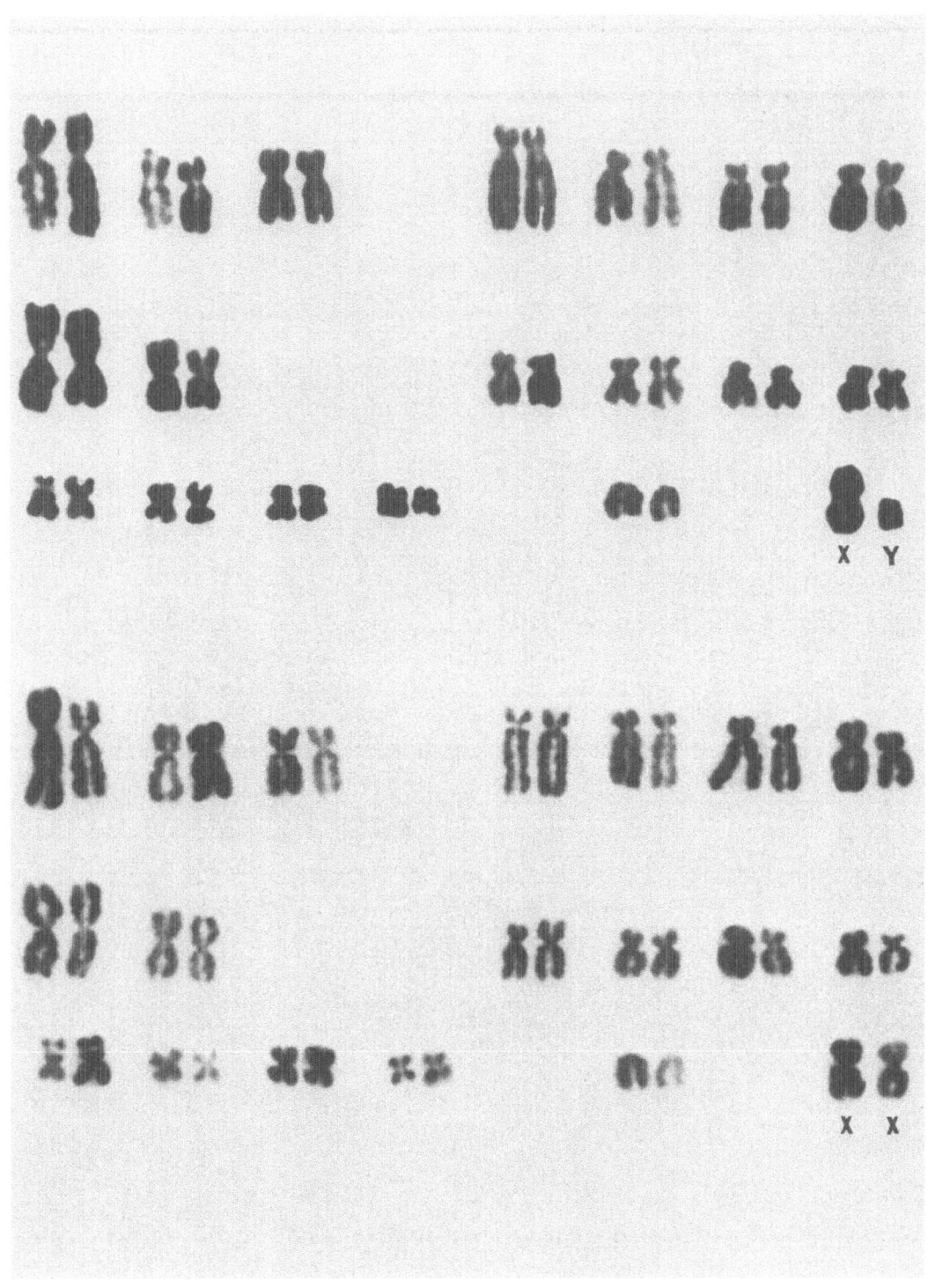

Order: CARNIVORA

Family: FELIDAE

Felis serval (Serval)

$2n = 38$

Felis serval (Serval)

2n=38

AUTOSOMES: 32 Metacentrics and submetacentrics
             4 Acrocentrics

SEX CHROMOSOMES: X Metacentric
                  Y Submetacentric

Skin biopsies of these two animals were kindly provided by Dr. C. Gray, National Zoological Park, Washington, D.C., USA. The karyotype is arranged following the San Juan Convention for Felidae (see Folio 31). The marker chromosome is the first in row five (E_1).

REFERENCES:

1) Wurster, D.H. and Benirschke, K.: Karyotypes of four more species of cats. Mammalian Chromosomes Newsletter 9:236, 1968.

2) Wurster, D.H. and Benirschke, K.: Comparative cytogenetic studies in the Order Carnivora. Chromosoma 24:336, 1968.

Order: CARNIVORA   Family: FELIDAE

Felis serval (Serval)

2n=38

Order: CARNIVORA

Family: FELIDAE

Felis viverrina (Fishing cat)
$2n = 38$

Volume 5, Folio 236, 1971

Order: CARNIVORA   Family: FELIDAE

Felis viverrina (Fishing cat)

2n=38

AUTOSOMES: 34 Metacentrics and submetacentrics
             2 Acrocentrics

SEX CHROMOSOMES: X Submetacentric
                   Y Submetacentric

     Skin biopsies of both animals were kindly provided by Dr. C. Gray,
National Zoological Park, Washington, D.C., USA. The karyotype is arranged
according to the San Juan Convention for Felidae (see Folio 31). The
marker chromosome (E_1) is the first element in the last row of each karyotype.

REFERENCES:

1) Wurster, D.H. and Benirschke, K.: The chromosomes of three species of
cats (Felis nigripes, F. bengalensis, F. viverrina). Mammalian Chromosomes
Newsletter 9:20, 1968.

2) Wurster, D.H. and Benirschke, K.: Comparative cytogenetic studies in
the Order Carnivora. Chromosoma 24:336, 1968.

3) Leyhausen, P. and Tonkin, B.A.: Comment on the karyotypes of the
leopard cat and the fishing cat. Mammalian Chromosomes Newsletter 9:78,
1968.

Order: CARNIVORA

Felis viverrina (Fishing cat)

2n=38

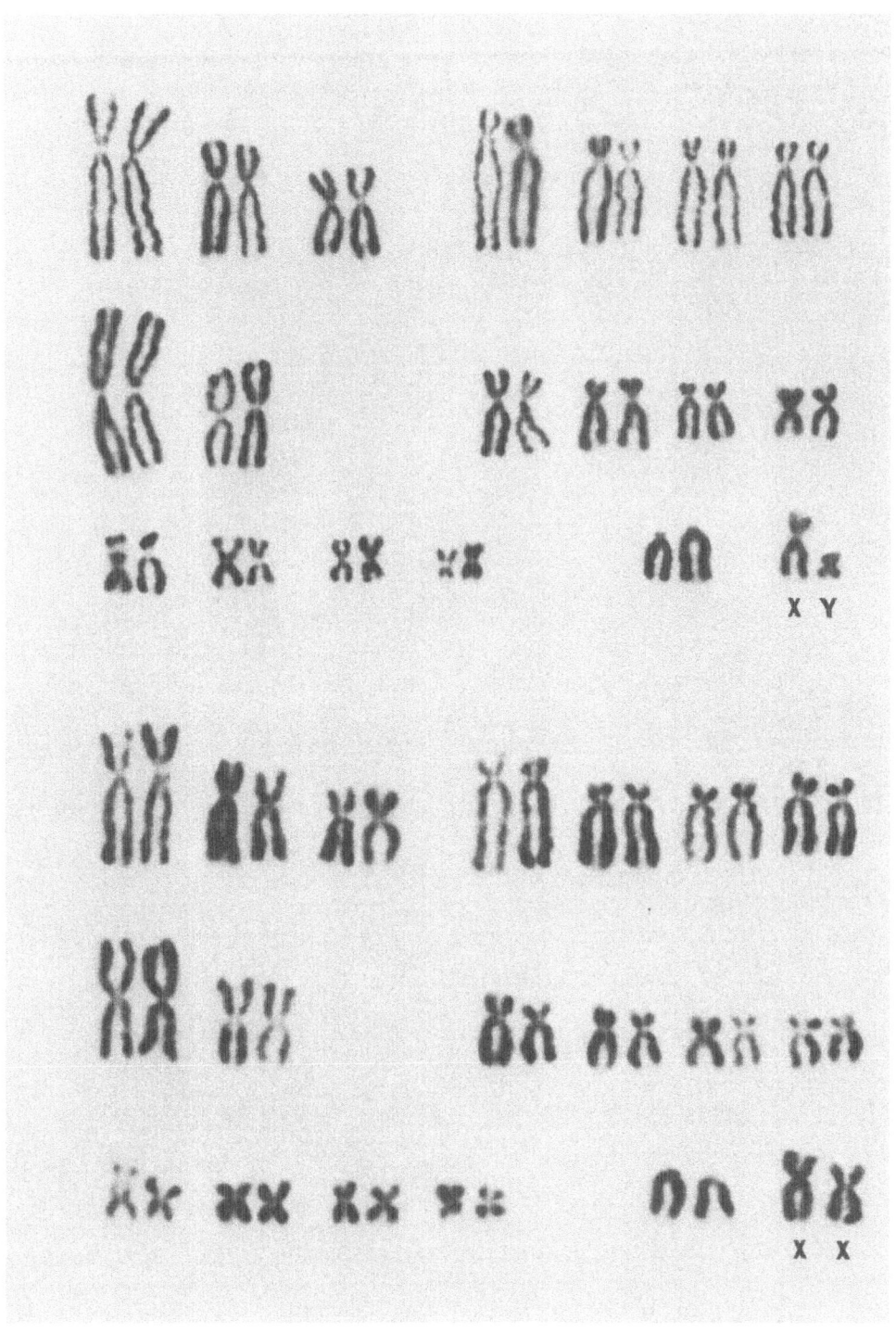

Order: CARNIVORA

Family: FELIDAE

Felis (Leopardus) wiedi(i) (Marguay cat)

2n = 36

Felis (Leopardus) wiedi(i) (Marguay cat)

2n=36

AUTOSOMES: 32 Metacentrics and submetacentrics
            2 Acrocentrics

SEX CHROMOSOMES: X Submetacentric
                  Y Submetacentric

A skin biopsy of a male specimen was kindly made available by
Dr. C. Gray of the National Zoological Park, Washington, D.C., USA. The
female karyotype was donated by Dr. R. Richart, New York, USA, and came
from an animal at Bronx Zoo. The karyotypes are arranged according to the
San Juan Convention for Felidae (see Folio 31). The marker is placed as
the first element in row five (E_1).

In the report of Hsu, Rearden and Luquette, the karyotype of F. wiedii
contained no acrocentric pairs. Leyhausen pointed out that the karyotype
matches that of F. tigrinus (little spotted cat). It is possible that the
specimen used in Hsu's study was misidentified and was actually F. tigrinus.
True F. wiedi has one pair of acrocentric autosomes.

REFERENCES:

1) Hsu, T.C.: Two species of cats with 36 chromosomes. Mammalian
Chromosomes Newsletter No. 8:4, 1962.

2) Hsu, T.C., Rearden, H. and Luquette, G.F.: Karyological studies of nine
species of Felidae. Amer. Naturalist 97:225, 1963.

3) Leyhausen, P.: The karyotypes of two cat species. Mammalian Chromosomes
Newsletter 8:287, 1967.

Felis (Leopardus) wiedi(i) (Marguay cat)

2n=36

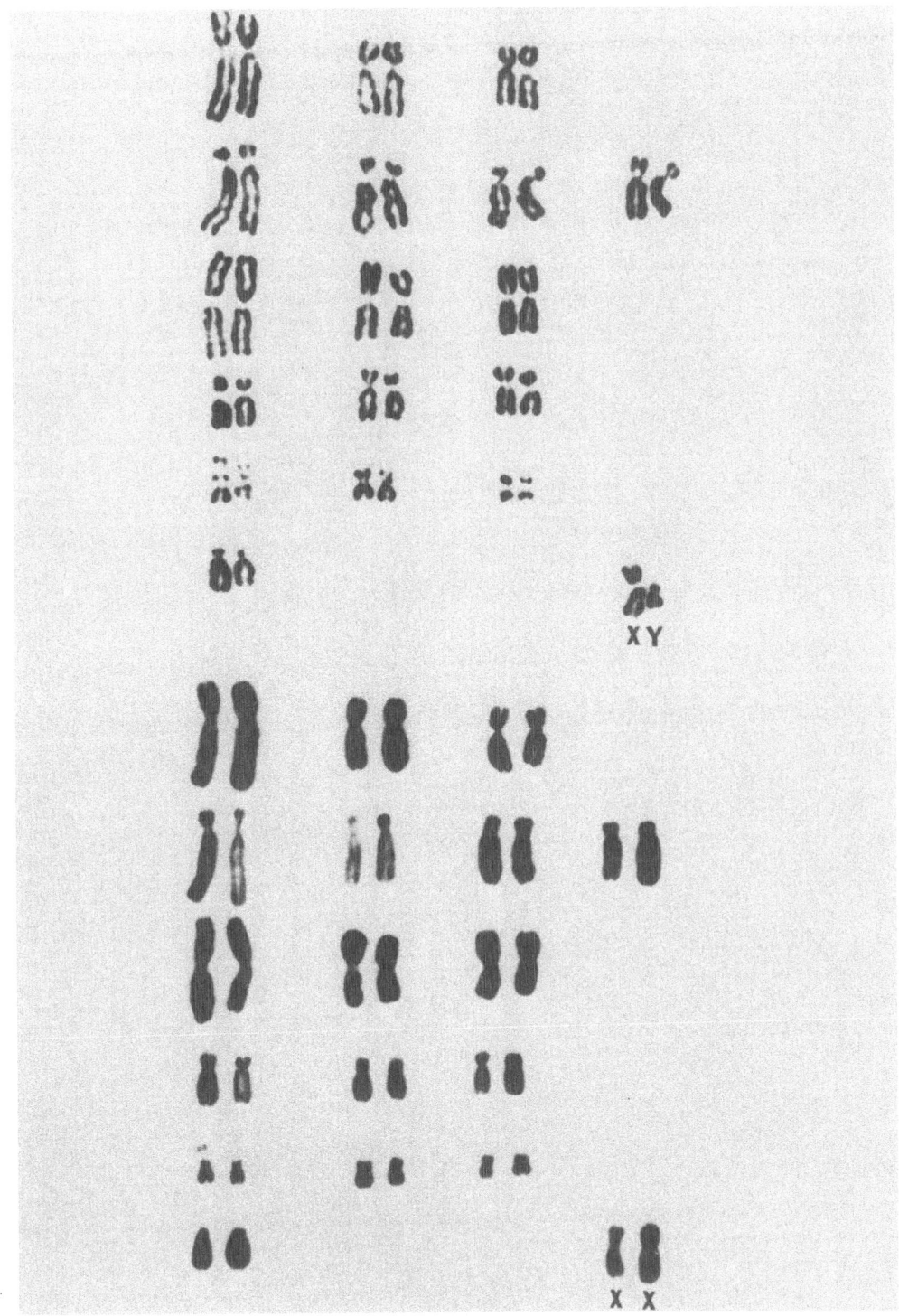

Order: TUBULIDENTATA

Family: ORYCTEROPODIDAE

Orycteropus afer (Aardvark)

$2n = 20$

Orycteropus afer (Aardvark)

2n=20

AUTOSOMES: 18 Metacentrics and submetacentrics

SEX CHROMOSOMES: X Metacentric
                   Y Metacentric

   The two largest autosomes often have narrow secondary constrictions which correspond to areas of late replication. The Y chromosome labels very late and heavily. The X chromosome constitutes approximately 5% of the haploid set. Dr. N. B. Atkin, Northwood, England, has measured the DNA content of cultured cells. The DNA content of these cells is approximately 1.6 x that of human lymphocytes.

   The male skin biopsy came from an animal at the National Zoological Park, Washington, D.C., USA, through the courtesy of Dr. C. Gray. The female karyotype came from a skin biopsy of an animal at Tacoma Zoo, Tacoma, Washington, USA, supplied by Dr. M. L. Johnson. A second female specimen from Crandon Park Zoo, Miami, Florida, USA, gave identical results.

REFERENCES:

1) Wurster, D. and Benirschke, K.: The chromosomes of the aardvark, Orycteropus afer. Mammalian Chromosomes Newsletter 11:34, 1970.

Order: TUBULIDENTATA   Family: ORYCTEROPODIDAE

Orycteropus afer (Aardvark)

2n=20

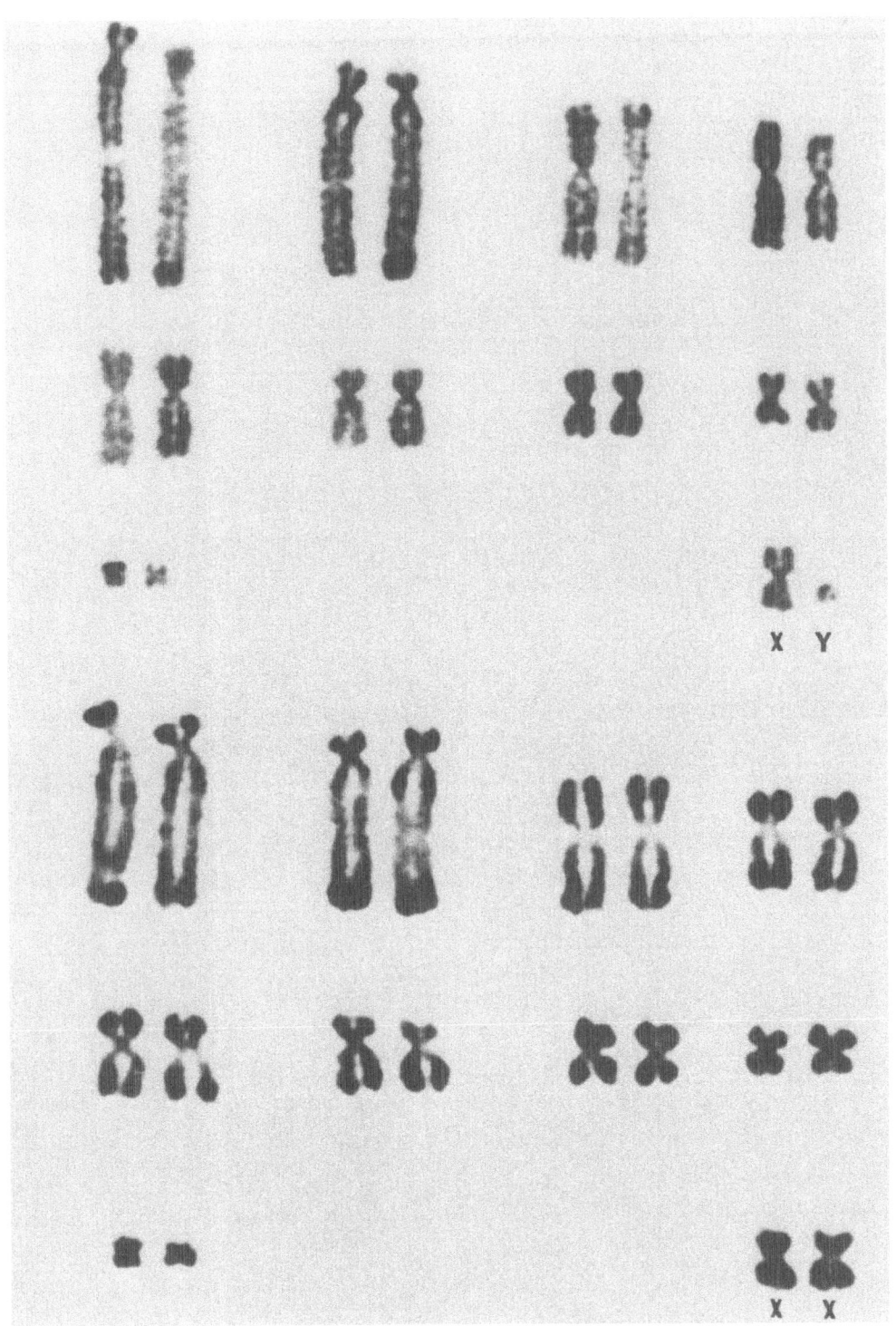

Order: PROBOSCIDEA

Family: ELEPHANTIDAE

Elephas maximus (Indian or Asian elephant)

2n = 56

Elephas maximus (Indian or Asian elephant)

2n=56

AUTOSOMES: 12 Metacentrics and submetacentrics
             42 Acrocentrics

SEX CHROMOSOMES: X Submetacentric
                  Y Submetacentric

    Among the biarmed autosomes, two pairs (the 2nd and the 3rd in row 1) may be classified as acrocentrics by some cytologists. In that case, the metacentric and submetacentric autosomes are reduced to 8. The Y chromosome may be called a subtelocentric by some authors.

    The karyotypes presented here are gifts of Drs. David A. Hungerford (Institute for Cancer Research, Philadelphia, Pennsylvania, USA) and R. L. Snyder (Penrose Research Laboratory, Zoological Society of Philadelphia, Pennsylvania, USA). The specimens were display animals of the Philadelphia Zoo. Another set of karyotypes donated to us by Dr. H. S. Norberg, Norway showed similar results. All were from cell cultures. We would like to thank Dr. Norberg for his kindness.

REFERENCES:

1) Hungerford, D. A., Chandra, H. S., Snyder, R. L. and Ulmer, F. A.: Chromosomes of three elephants, two Asian (Elephas maximus) and one African (Loxodonta africana). Cytogenetics 5:243, 1966.

2) Norberg, H. S.: The chromosomes of the Indian female elephant (Elephas indicus syn. E. maximus L.). Hereditas 63:279, 1969.

Elephas maximus (Indian or Asian elephant)

2n=56

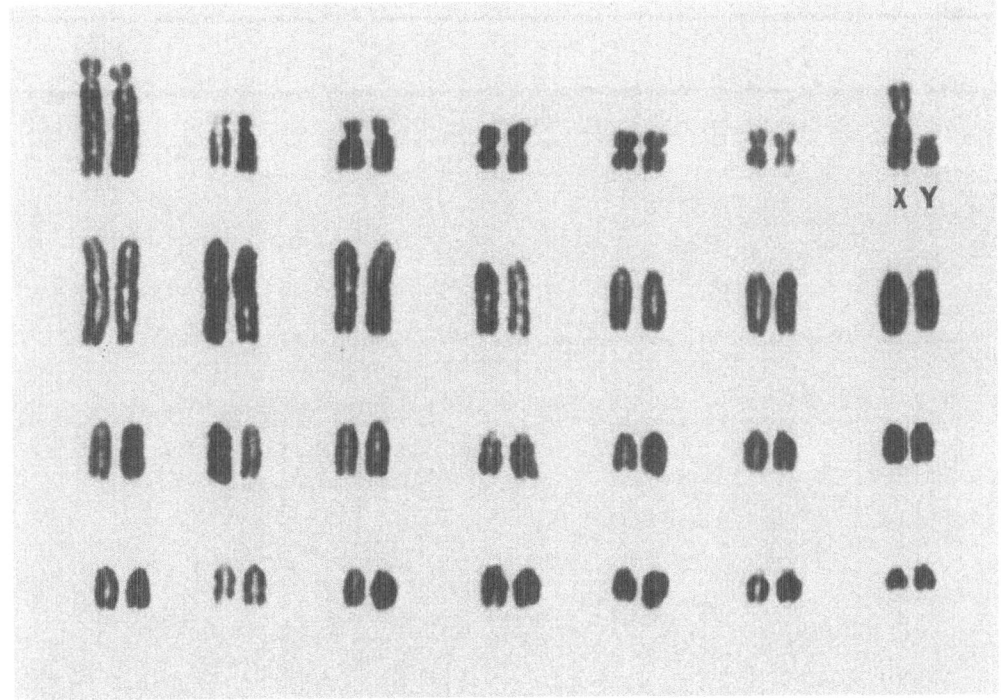

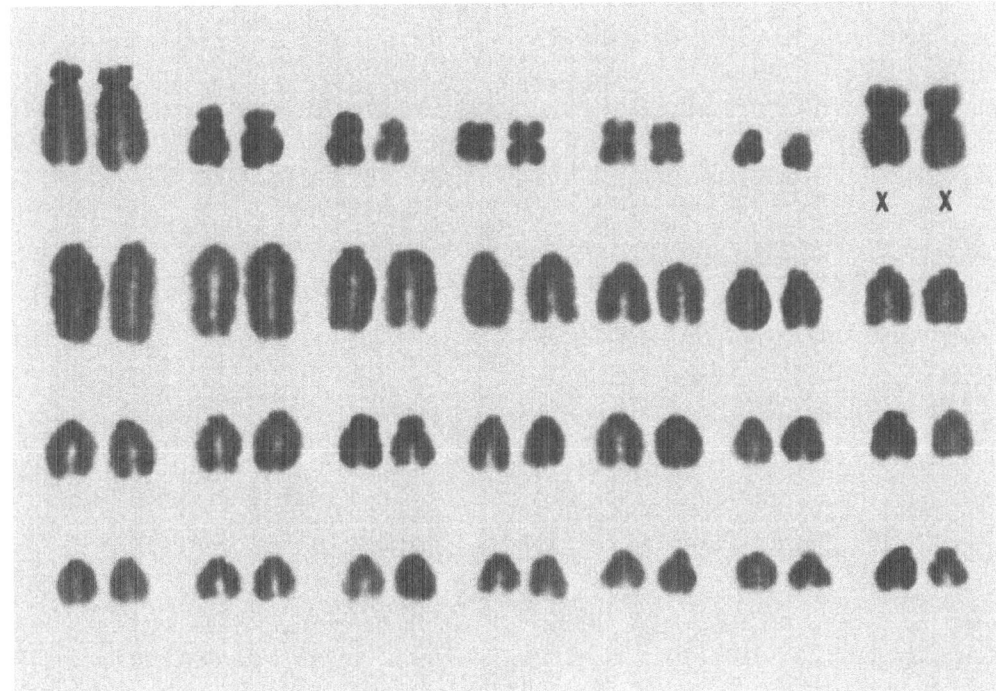

Order: PERISSODACTYLA

Family: EQUIDAE

Equus quagga (Burchell's zebra, Plains zebra)

2n = 44

Equus quagga (Burchell's zebra, Plains zebra)

2n=44

AUTOSOMES: 34 Metacentrics and submetacentrics
             8 Acrocentrics

SEX CHROMOSOMES: X Metacentric
                   Y Minute ?acrocentric

     Some of the smaller autosomes are difficult to classify, hence the
species is often considered to possess only 6 acrocentric autosomes. Also,
the Y is so small that it is not certain to be acrocentric.

     Several different subspecies of this common zebra have been studied,
all with similar results. The male karyotype shown here is a Damara zebra
(E. q. antiquorum), the female is of the rare "maneless" mutant
(E. q. boehmi). Skin biopsies were kindly provided by Dr. H. Heck,
Catskill Game Farm, New York, USA, who also identified the species.

REFERENCES:

1) Benirschke, K. and McFeeley, R.A.: The chromosomes of the Grant zebra,
Equus quagga boehmi. Mammalian Chromosomes Newsletter No. 10:82, 1963.

2) Mutton, D.E., King, J.M. and Hamerton, J.L.: Chromosome studies in the
genus Equus. Mammalian Chromosomes Newsletter No. 13:7, 1964.

3) Benirschke, K. and Malouf, N.: Chromosome studies of Equidae. In
Equus, Vols. 1 & 2 (H. Dathe, ed.), Tierpark Berlin, 253, 1967.

4) King, J.M., Short, R.V., Mutton, D.E. and Hamerton, J.L.: The
reproductive physiology of male zebra-horse and zebra-donkey hybrids.
J. Reprod. Fertil. 9:391, 1965.

Equus quagga (Burchell's zebra, Plains zebra)

2n=44

Order: ARTIODACTYLA

Family: CERVIDAE

Elaphurus davidianus (Pere David's deer)

$2n = 68$

Volume 5, Folio 241, 1971

Order: ARTIODACTYLA   Family: CERVIDAE

<u>Elaphurus</u> <u>davidianus</u> (Père David's deer)

2n=68

AUTOSOMES: 2 Metacentrics
             64 Acrocentrics

SEX CHROMOSOMES: X Acrocentric
                  Y Metacentric

     Skin biopsies were kindly made available of a male by Dr. C. Gray,
National Zoological Park, Washington, D.C., USA, and of a female by
Dr. H. Heck, Catskill Game Farm, New York, USA.

Elaphurus davidianus (Père David's deer)

2n=68

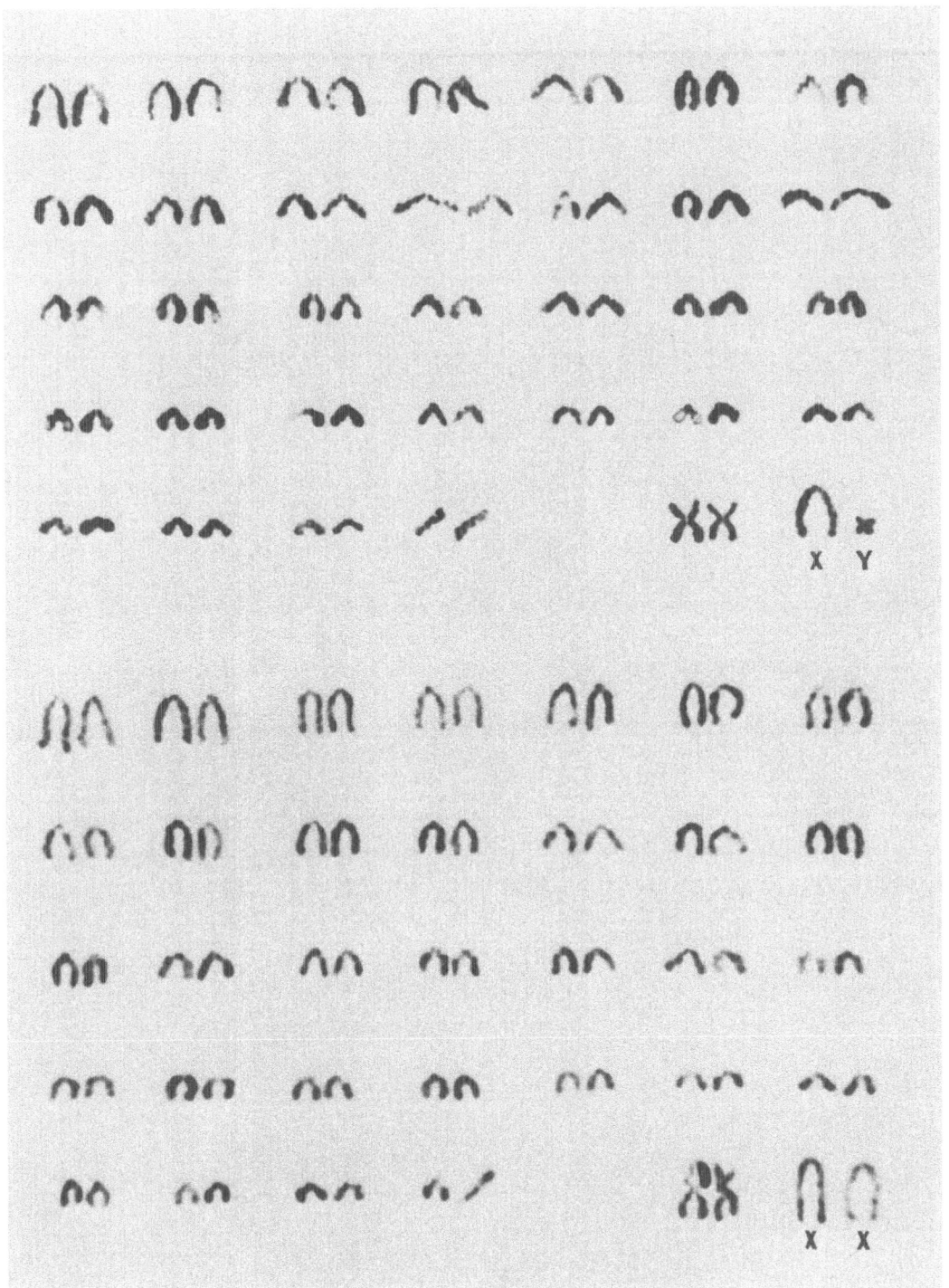

Order: ARTIODACTYLA

Family: BOVIDAE

Cephalophus silvicultor (Yellow-backed duiker)
$2n = 60$

Order: ARTIODACTYLA   Family: BOVIDAE

<u>Cephalophus silvicultor</u> (Yellow-backed duiker)

2n=60

AUTOSOMES: 58 Acrocentrics

SEX CHROMOSOMES: X Submetacentric
                  Y Acrocentric

    Blood cultures of one male and one female animal were used for the preparation of these karyotypes. They were kindly donated by Dr. W. L. Hard, Omaha, Nebraska, USA, and came from animals at the Henry Doorley Zoo, Omaha.

    While the identification of the X is unequivocal, the Y is not positively identified. Several other duikers have similar karyotypes as discussed by Hard.

REFERENCES:

1) Hard, W.L.: The chromosomes of duikers. Mammalian Chromosomes Newsletter <u>10</u>:216, 1969.

Cephalophus silvicultor (Yellow-backed duiker)

2n=60

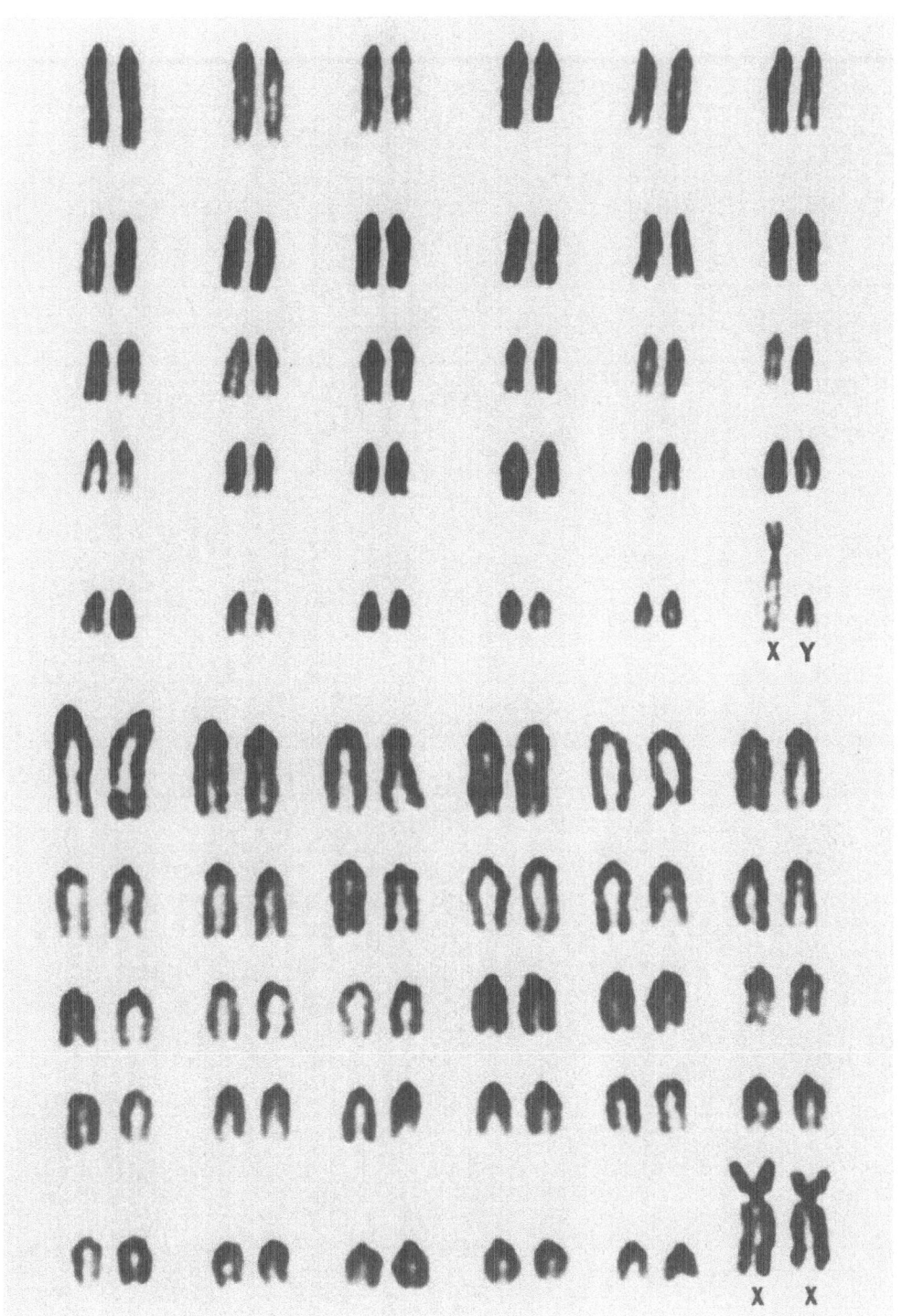

Order: ARTIODACTYLA

Family: BOVIDAE

Rupicapra rupicapra (Chamois)
$2n = 58$

Rupicapra rupicapra (Chamois)

2n=58

AUTOSOMES: 2 Metacentrics
             54 Acrocentrics

SEX CHROMOSOMES: X Acrocentric
                  Y Metacentric

     These karyotypes, prepared from kidney cultures, were kindly donated
by Dr. A. Gropp, Bonn, Germany. Three animals were studied from Valais,
Switzerland. The X was identified by autoradiography. The paracentromeric
region of the large metacentric is late replicating.

REFERENCES:

1) Gropp, A., Giers, D. and Wandeler, A.: The karyotype of the chamois
(Rupicapra rupicapra, Bovidae). Mammalian Chromosomes Newsletter 10:19,
1969.

2) Gropp, A., Giers, D., Fernandez-Donoso, R., Tiepolo, L. and Fraccaro,
M.: The chromosomes of the chamois (tribe Rupicaprini Simpson).
Cytogenetics 9:1, 1970.

Rupicapra rupicapra (Chamois)

2n=58

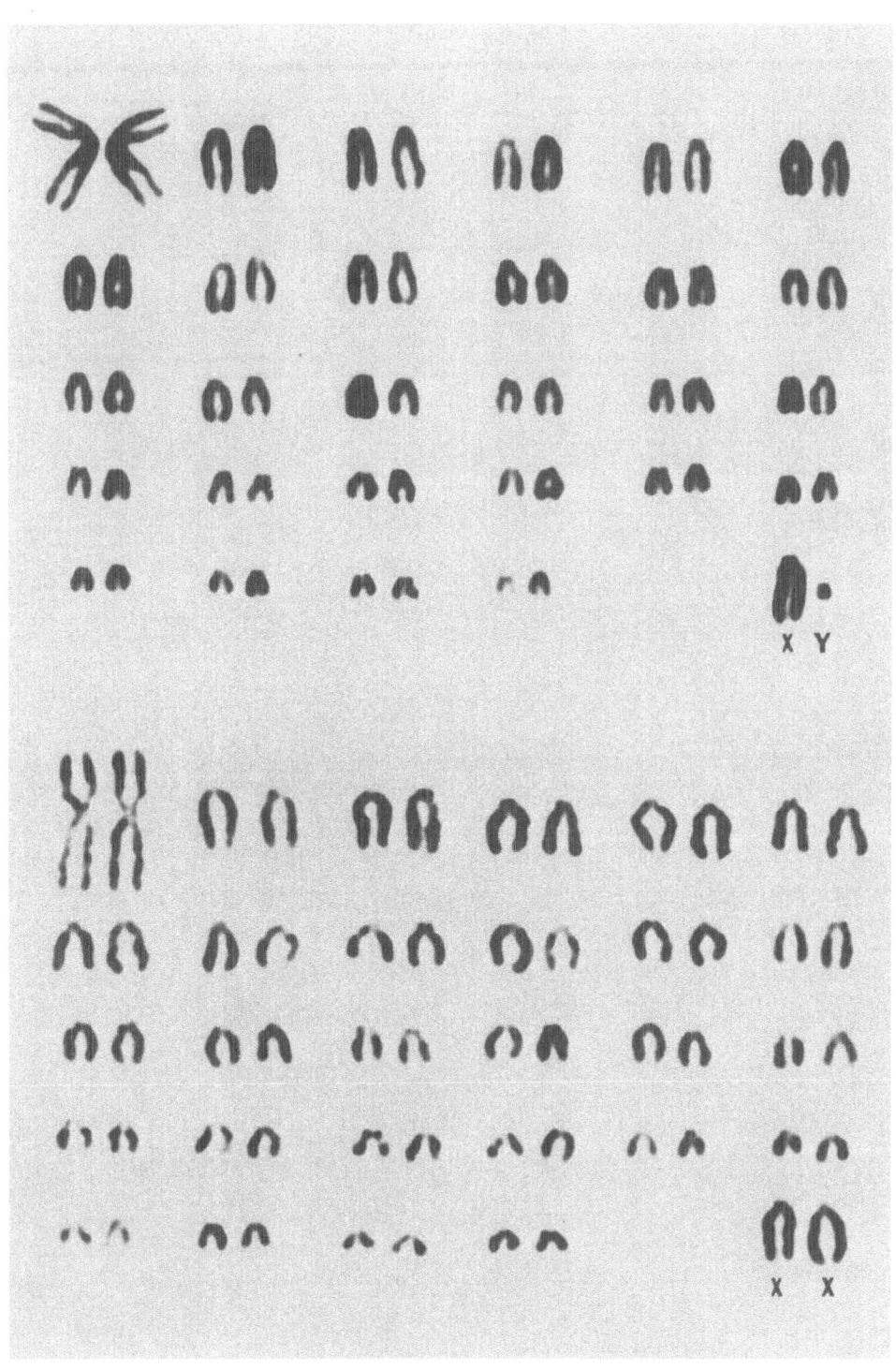

Order: PRIMATES

Lemur fulvus fulvus (Red-fronted lemur)

Family: LEMURIDAE
2n = 48

<u>Lemur</u> <u>fulvus</u> <u>fulvus</u> (Red-fronted lemur)

2n=48

AUTOSOMES: 16 Metacentrics and submetacentrics
             30 Acrocentrics

SEX CHROMOSOMES: X Acrocentric
                   Y Minute

    Among the large metacentrics and submetacentrics roughly two groups of 4 pairs each can be recognized, one with the two arms more or less equal and the other distinctly unequal. One of the submetacentric pairs is unusually long. All the acrocentrics are relatively small, three pairs of which appear to bear secondary constriction near the centromere.

    The sex chromosome can be unequivocally identified, the X being the longest acrocentric of the complement, and the Y, the smallest.

    The karyotypes are gifts of Dr. Ernest H.Y. Chu, Biology Division, Oak Ridge National Laboratory, Oak Ridge, Tennessee, USA. The specimens were collected by Dr. John Buettner-Janusch in Madagascar.

REFERENCES:

1) Chu, E.H.Y. and Swomley, B.A.: Chromosomes of lemurine lemurs. Science 133:1925, 1961.

2) Bender, M.A. and Chu, E.H.Y.: The chromosomes of Primates. In "Evolutionary and Genetic Biology of Primates", Vol. 1 (J. Buettner-Janusch, ed.), Academic Press, N.Y., 1963.

3) Egozcue, J.: Chromosome variability in the Lemuridae. Amer. J. Phys. Anthrop. 26:341, 1967.

4) Egozcue, J.: Primates. In "Comparative Mammalian Cytogenetics" (Benirschke, K., ed.), Springer-Verlag, N.Y., 1969.

Lemur fulvus fulvus (Red-fronted lemur)

2n=48

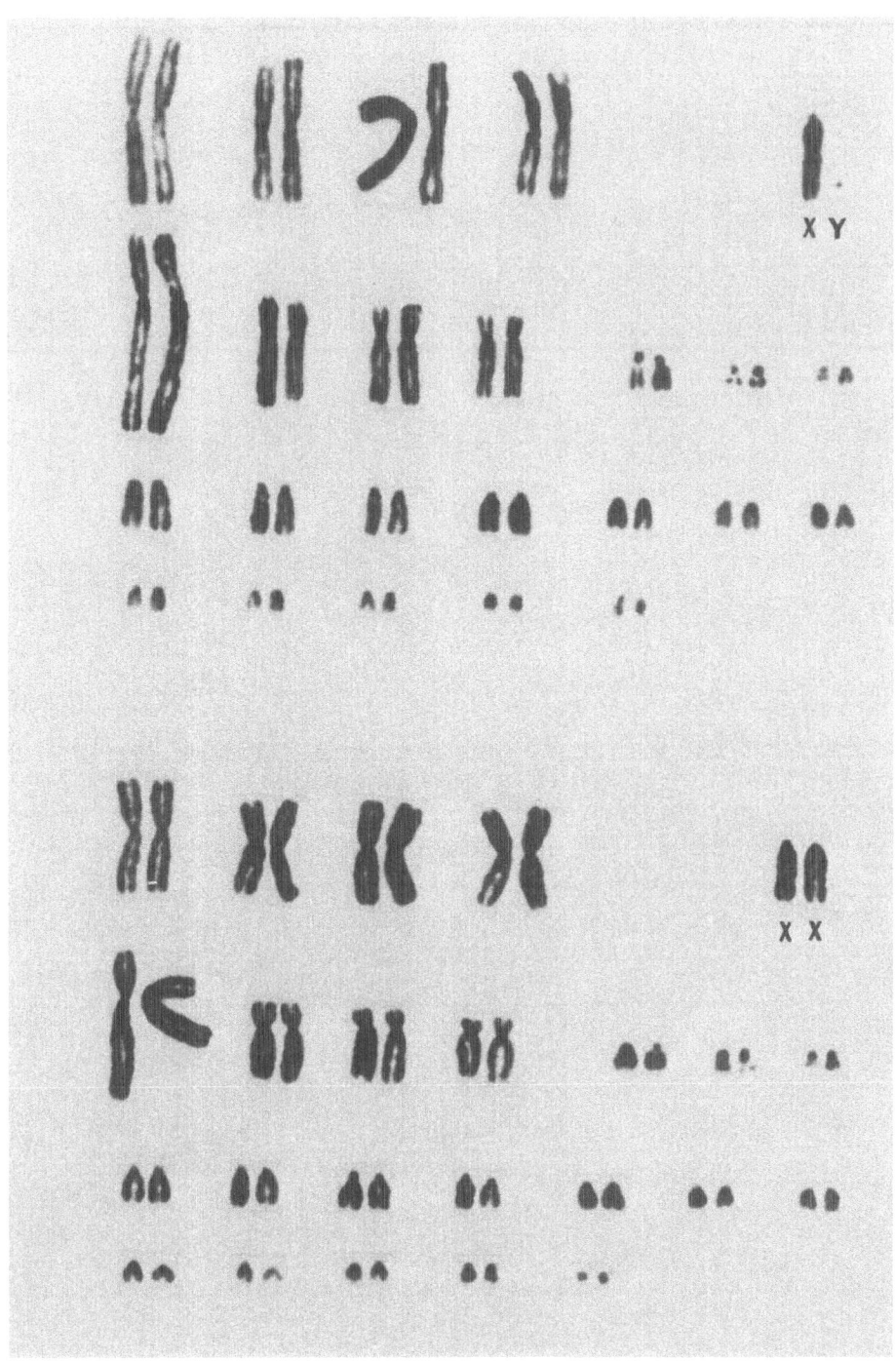

Order: PRIMATES

Leontocebus rosalia (Golden marmoset)

Family: CALLITHRICIDAE
2n = 46

<u>Leontocebus</u> <u>rosalia</u> (Golden marmoset)

2n=46

AUTOSOMES: 42 Metacentrics, submetacentrics and subtelocentrics
            2 Acrocentrics

SEX CHROMOSOMES: X Submetacentric
                  Y Metacentric

    The short arms of several autosomes are of sufficient length to place
the elements into the group of subtelocentrics or submetacentrics;
division is arbitrary and depends on the excellence of preparation. Bone
marrows usually yield less favorable results. Male/female blood chimerism
is found in many animals. The male karyotype came from a skin culture, the
female karyotype from a bone marrow preparation. Only one pair is considered
to be truly acrocentric and is placed here before the sex chromosomes.

REFERENCES:

1) Benirschke, K., Anderson, J.M. and Brownhill, L.E.: Marrow chimerism
in marmosets. Science <u>138</u>:513, 1962.

2) Benirschke, K. and Brownhill, L.E.: Further observations on marrow
chimerism in marmosets. Cytogenetics <u>1</u>:245, 1962.

3) Egozcue, J.: Primates. In <u>Comparative Mammalian Cytogenetics</u>
(Benirschke, K., ed.), Springer-Verlag, New York, 1969.

4) Hsu, T.C. and Hampton, S.H.: Chromosomes of Callithricidae with
special reference to an XX/'XO' sex chromosome system in <u>Callimico</u> <u>goeldii</u>
(Thomas, 1904). Folia Primatologica (in press).

Leontocebus rosalia (Golden marmoset)

2n=46

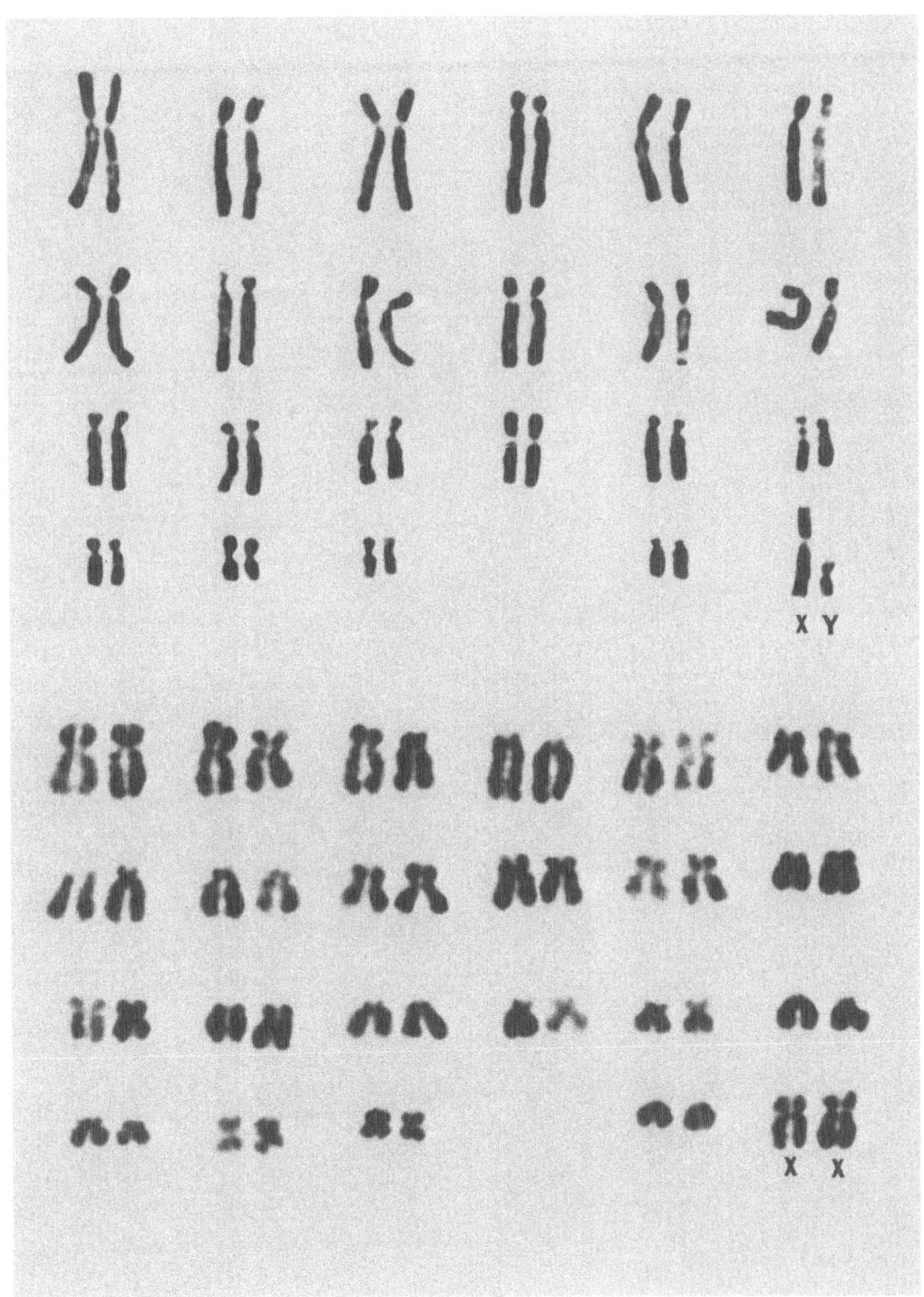

Order: PRIMATES

Family: CERCOPITHECIDAE

Cercocebus torquatus (Red crowned mangabey)

$2n = 42$

Cercocebus torquatus (Red crowned mangabey)

2n=42

AUTOSOMES: 40 Metacentrics and submetacentrics

SEX CHROMOSOMES: X Submetacentric
                  Y Metacentric

     Skin biopsies of a male and a female animal were kindly made
available by Dr. C. Gray, National Zoological Park, Washington, D.C., USA.
The Y chromosome is easily identified as the smallest element. The pair
of marker chromosomes with secondary paracentromeric constrictions is
placed as the last autosomes.

REFERENCES:

1) Chu, E.H.Y. and Giles, N.M.: A study of primate chromosome complements.
Amer. Natur. 41:273, 1957.

2) Bender, M.A. and Mettler, L.E.: Chromosome studies of primates.
Science 128:186, 1958.

3) Chiarelli, B.: Comparative morphometric analysis of the primate
chromosomes. II. The chromosomes of the genera Macaca, Papio, Theropithecus,
and Cercocebus. Caryologia 15:401, 1962.

4) Klinger, H.P.: The somatic chromosomes of some primates (Tupaia glis,
Nycticebus coucang, Tarsius bancanus, Cercocebus aterrimus, Symphalangus
syndactylus). Cytogenetics 2:140, 1963.

5) Chiarelli, B.: Marked chromosome in catarrhine monkeys. Folia primat.
4:74, 1966.

6) Chiarelli, B.: La morfologia del cromosoma "Y" delle differenti specie
di primati. Rivista Antropol. 54:3, 1967.

7) Chiarelli, B. and Vaccarino, C.: Cariologia ed evoluzione nel genere
Cercopithecus. Atti Ass. Genet. It., Pavia 9:328, 1964.

8) Chiarelli, B.: Caryology and taxonomy of the catarrhine monkeys.
Amer. J. Phys. Anthropol. 24:155, 1966.

9) Wurster, D. and Benirschke, K.: Chromosomes of some primates.
Mammalian Chromosomes Newsletter 10:3, 1969.

Order: PRIMATES

Cercocebus torquatus (Red crowned mangabey)

2n=42

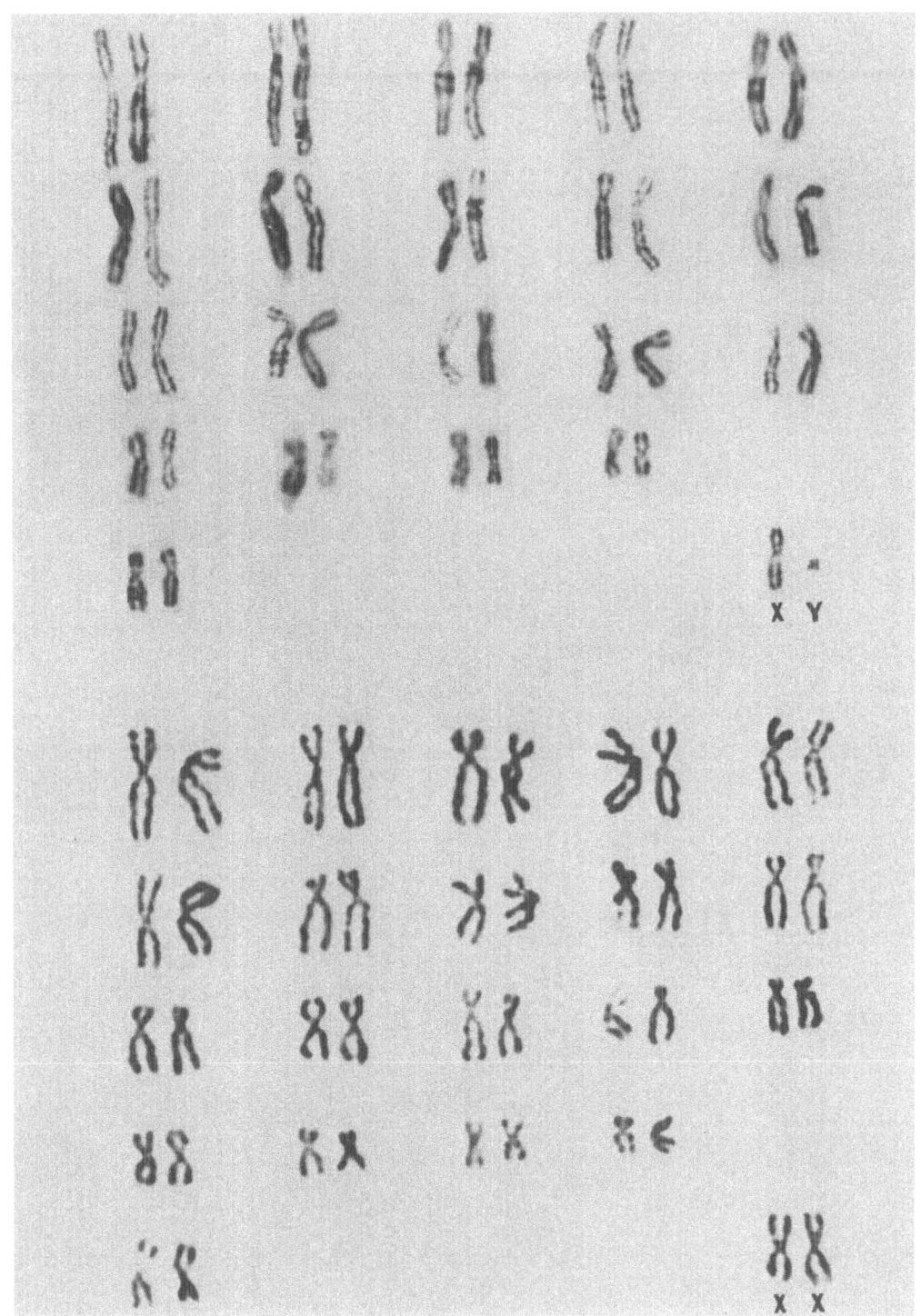

Order: PRIMATES

Family: CERCOPITHECIDAE

Cercopithecus cephus (Moustached guenon)
$2n = 66$

Cercopithecus cephus (Moustached guenon)

2n=66

AUTOSOMES: 44 Metacentrics, submetacentrics and subtelocentrics
            20 Acrocentrics

SEX CHROMOSOMES: X Submetacentric
                  Y Submetacentric

Some subtelocentric autosomes could easily be regarded as acrocentrics. The marker autosome is placed before the sex chromosomes. It is acrocentric and has secondary constrictions near the centromere. Skin biopsies of one male and three female animals yielded similar results. They were kindly provided by Dr. C. Gray, National Zoological Park, Washington, D.C., USA.

REFERENCES:

1) Wurster, D. and Benirschke, K.: Chromosomes of some primates. Mammalian Chromosomes Newsletter 10:3, 1969.

2) Chiarelli, B.: Primi resultati di richerche di genetica e cariologia comparata in primati e loro interese evolutivo. Rivista di Antropol. 50: 87, 1963.

3) Egozcue, J.: Primates. In Comparative Mammalian Cytogenetics (Benirschke, K., ed.), Springer-Verlag, New York, 1969.

4) Chiarelli, B.: Survey of the caryology of the genus Cercopithecus. Mammalian Chromosomes Newsletter No. 22, 209, 1966.

5) Chiarelli, B. and Vaccarino, C.: Cariologia ed evoluzione nel genere Cercopithecus. Atti Ass. Genet. It., Pavia 9:328, 1964.

6) Chiarelli, B.: Chromosome polymorphism in the species of the genus Cercopithecus. Cytologia 33:1, 1968.

Cercopithecus cephus (Moustached guenon)

2n=66

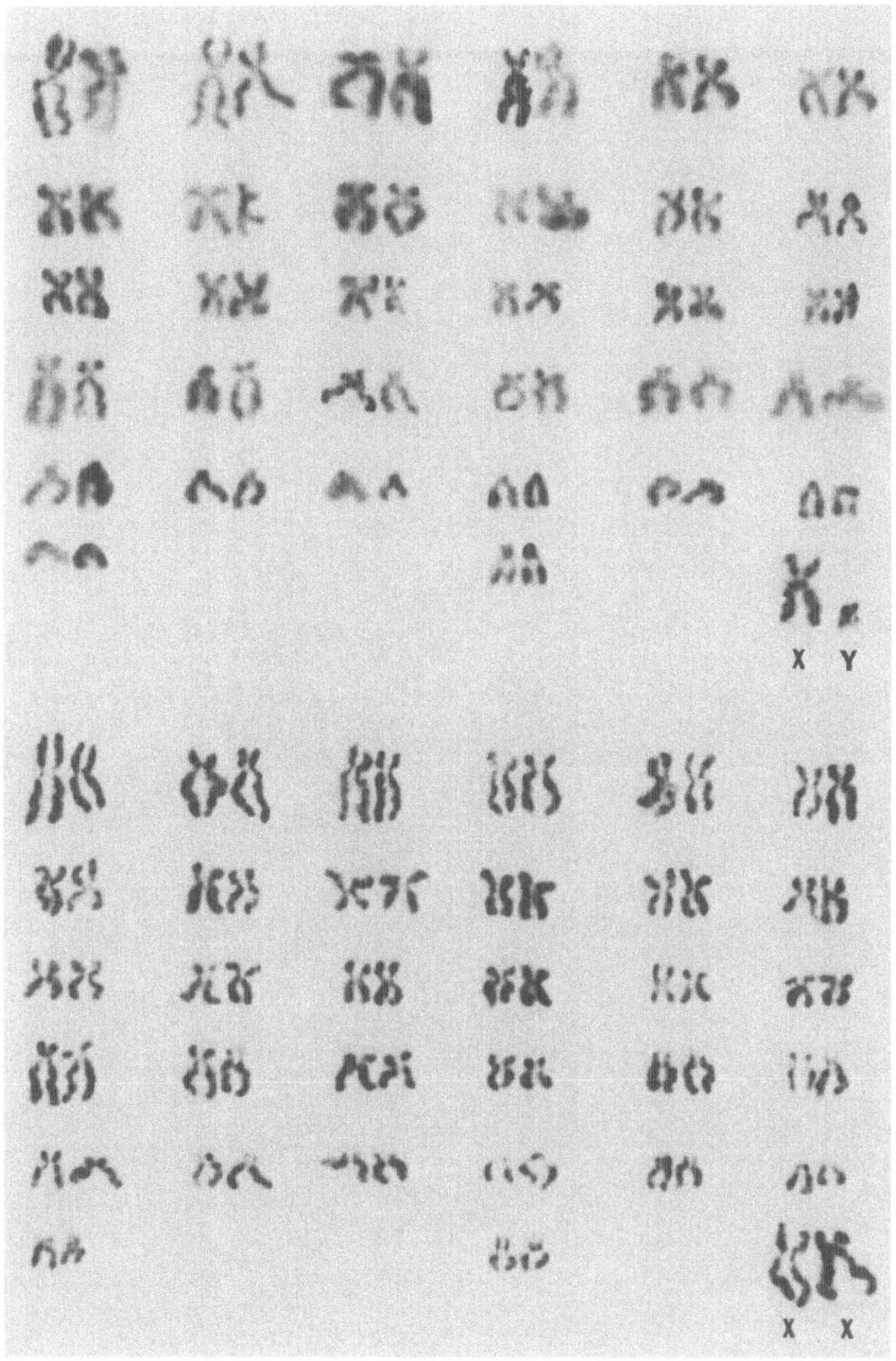

Order: PRIMATES

Family: CERCOPITHECIDAE

Miopithecus (Cercopithecus) talapoin (Talapoin monkey; ozem)
2n = 54

Order: PRIMATES   Family: CERCOPITHECIDAE

Miopithecus (Cercopithecus) talapoin (Talapoin monkey; ozem)

2n=54

AUTOSOMES: 38 Metacentrics and submetacentrics
             14 Acrocentrics

SEX CHROMOSOMES: X Submetacentric
                  Y Submetacentric, minute

These karyotypes came from tissue cultures of skin biopsies and were donated by Dr. T. C. Jones, New England Regional Primate Research Center, Southborough, Massachusetts, USA. They were prepared in collaboration with Dr. R. Cooper, San Diego Zoo, California, USA. The arrangement into four groups of autosomes is that suggested by Dr. Jones. The acrocentrics shown above the sex chromosomes are the marker elements with a secondary constriction near the centromere. The Y has also been described as being acrocentric.

REFERENCES:

1) Chiarelli, B. and Vaccarino, C.: Cariologia ed evoluzione nel genere Cercopithecus. Atti Ass. Genet. It., Pavia 9:328, 1964.

2) Chiarelli, B.: Caryology and taxonomy of the catarrhine monkeys. Amer. J. Phys. Anthrop. 24:155, 1966.

3) Barberis, L.: The idiograms of three species of the genus Cercopithecus, C. talapoin, C. aethiops and C. mona. In Progress in Primatology (Starck, D., Schneider, R. and Kuhn, H.J., eds.), Gustav Fischer, Stuttgart, 1967.

4) Chiarelli, B.: Chromosome polymorphism in the species of the genus Cercopithecus. Cytologia 33:1, 1968.

5) Chiarelli, B.: Some new data on the chromosomes of Catarrhina. Experientia 18:405, 1962.

6) Egozcue, J.: Primates. In Comparative Mammalian Cytogenetics (Benirschke, K., ed.), Springer-Verlag, New York, 1969.

Miopithecus (Cercopithecus) talapoin (Talapoin monkey; ozem)

2n=54

Order: PRIMATES

Family: CERCOPITHECIDAE

Presbytis obscurus (Dusky langur)

2n = 48

Order: PRIMATES   Family: CERCOPITHECIDAE

Presbytis obscurus (Dusky langur)

2n=44

AUTOSOMES: 40 Metacentrics and submetacentrics
             2 Acrocentrics

SEX CHROMOSOMES: X Submetacentric
                  Y Submetacentric

Skin biopsies of a male and female animal were kindly made available by Dr. C. Gray, National Zoological Park, Washington, D.C., USA. The karyotype has previously (1) been described erroneously as belonging to P. phayrei and we are grateful to Dr. C. A. Hill, San Diego Zoo, California, USA., for clarifying the error.

Pairing of autosomes is relatively arbitrary except for the smallest metacentrics, the acrocentric pair, and the marker pair which is placed as the last autosomal pair. At least one metacentric has very small satellites.

REFERENCES:

1) Wurster, D. and Benirschke, K.: Chromosomes of some primates. Mammalian Chromosomes Newsletter 10:3, 1969.

2) Chiarelli, B.: Cariologia ed evoluzione nel genere Cercopithecus. Atti Ass. Genet. It., Pavia 9:328, 1964.

3) Chiarelli, B.: Caryology and taxonomy of the catarrhine monkeys. Amer. J. Phys. Anthropol. 24:155, 1966.

4) Chiarelli, B.: La morfologia del cromosome "Y" delle differenti specie di primati. Rivista Antropol. 54:3, 1967.

5) Chiarelli, B.: Comparative morphometric analysis of primate chromosomes. III. The chromosomes of the genera Hylobates, Colobus and Presbytis. Caryologia 16:637, 1963.

<u>Presbytis</u> <u>obscurus</u> (Dusky langur)

2n=44

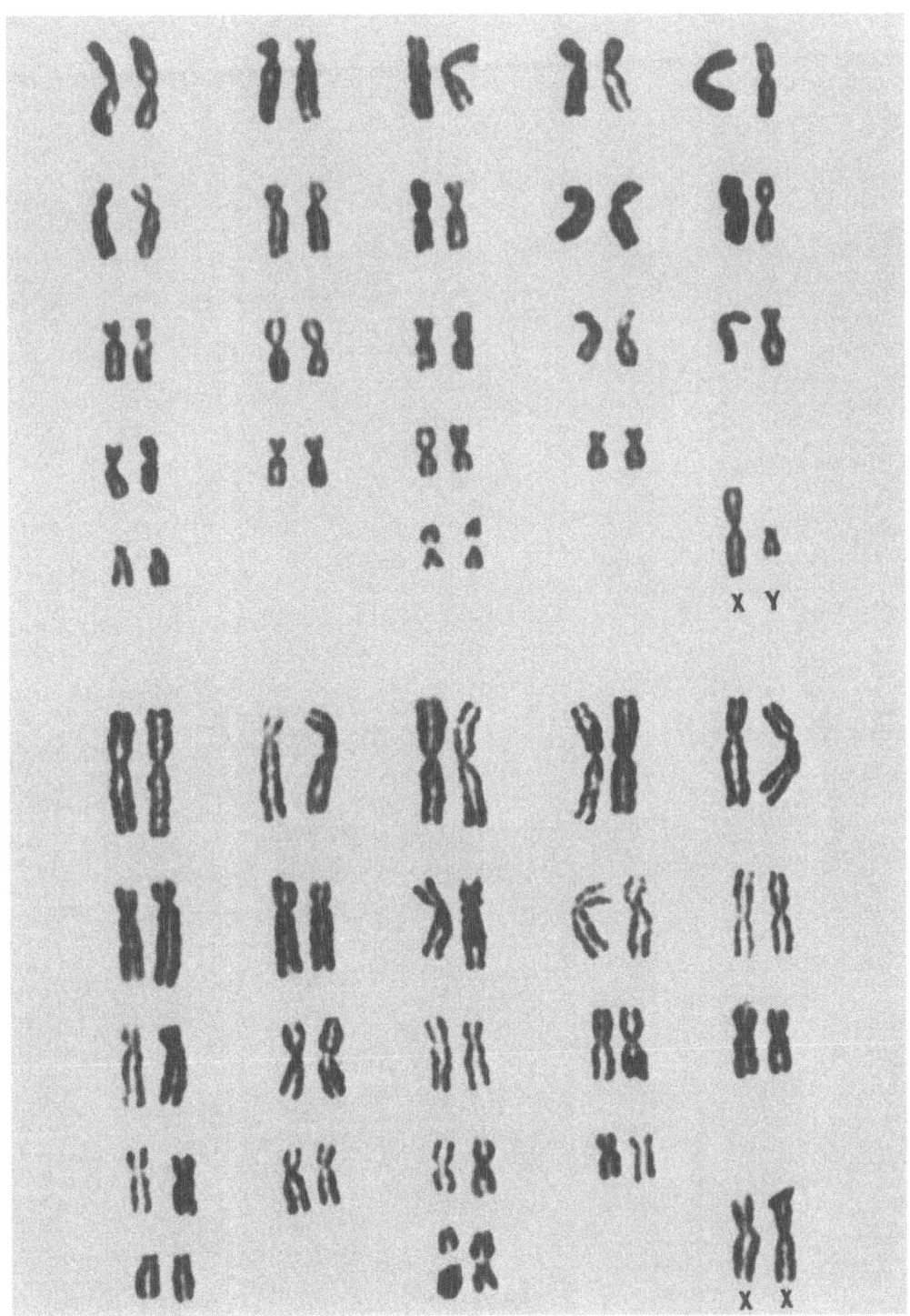

Order: PRIMATES

Family: PONGIDAE

Pongo pygmaeus (Orangutan)
$2n = 44$

Pongo pygmaeus (Orangutan)

2n=48

AUTOSOMES: 28 Metacentrics or submetacentrics
             18 Acrocentrics

SEX CHROMOSOMES: X Submetacentric
                  Y Acrocentric or submetacentric

    Skin biopsies of one male and one female were kindly made available
by Dr. C. Gray, National Zoological Park, Washington, D.C., USA. Lymphocyte
cultures were obtained of the pair of fraternal twins at the Zoo in Seattle,
Washington, USA, through the courtesy of Dr. D. Smith. Skin biopsy of a male
newborn came from the San Diego Zoo. All gave similar results; no chimerism
was found in the twins. The karyotypes displayed here came from skin
cultures, the male from "Archie", the female from "Susie" at Washington, D.C.
Faint satellites can be seen on some acrocentrics.

REFERENCES:

1) Chiarelli, B.: Chromosomes of the Orang-Utan (Pongo pygmaeus). Nature
192:825, 1961.

2) Hamerton, J.L., Klinger, H.P., Mutton, D.E. and Lang, E.M.: The somatic
chromosomes of the Hominoidea. Cytogenetics 2:240, 1963.

3) Klinger, H.P., Hamerton, J.L., Mutton, D. and Lang, E.M.: The chromosomes
of the Hominoidea. In Classification and Human Evolution (Washburn, S.L., ed.),
Aldine Publishing Co., Chicago, 1963.

4) Egozcue, J.: Primates. In Comparative Mammalian Cytogenetics
(Benirschke, K., ed.), Springer-Verlag, New York, 1969.

5) Chiarelli, B.: Comparative morphometric analysis of the primate
chromosomes. I. The chromosomes of the anthropoid apes and of man.
Caryologia 15:99, 1962.

6) Chiarelli, B.: La morfologia del cromosoma "Y" delle differenti specie
di primati. Rivista Antropol. 54:3, 1967.

Pongo pygmaeus (Orangutan)

2n=48

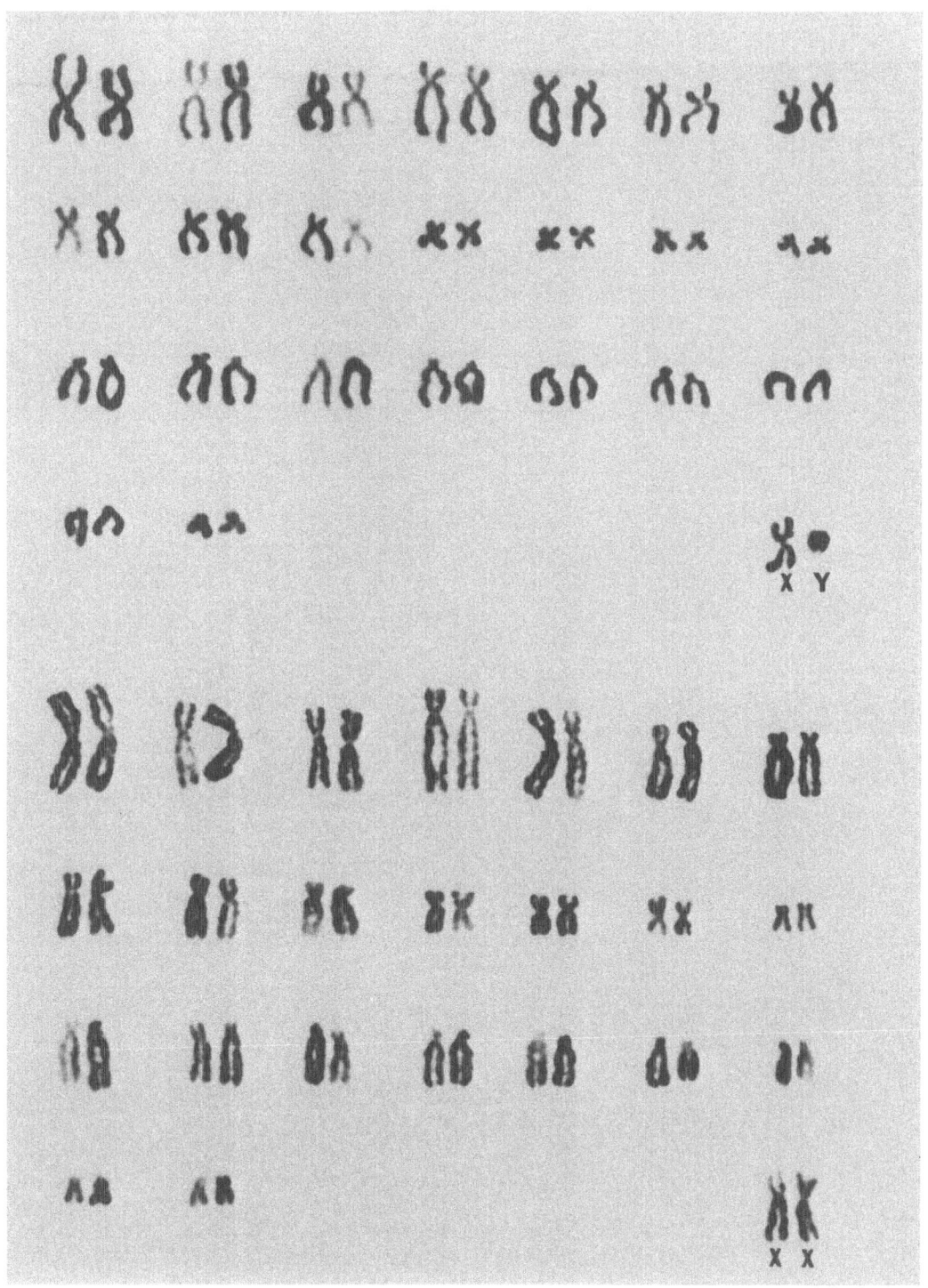

Cumulative Index (Volumes 1 to 5)

Vernacular Names

Cumulative Index (Volumes 1 to 5)

Technical Names